Active Galactic Nuclei

Fueling and feedback

AMERICAN ASTRONOMICAL SOCIETY IOP | ebooks™

AAS Editor in Chief

Ethan Vishniac, Johns Hopkins University, Maryland, USA

About the program:

AAS-IOP Astronomy ebooks is the official book program of the American Astronomical Society (AAS) and aims to share in depth the most fascinating areas of astronomy, astrophysics, solar physics, and planetary science. The program includes publications in the following topics:

Books in the program range in level from short introductory texts on fast-moving areas, graduate and upper-level undergraduate textbooks, research monographs, and practical handbooks.

For a complete list of published and forthcoming titles, please visit iopscience.org/books/aas.

About the American Astronomical Society

The American Astronomical Society (aas.org), established 1899, is the major organization of professional astronomers in North America. The membership (~7,000) also includes physicists, mathematicians, geologists, engineers, and others whose research interests lie within the broad spectrum of subjects now comprising the contemporary astronomical sciences. The mission of the Society is to enhance and share humanity's scientific understanding of the universe.

Editorial Advisory Board

Active Galactic Nuclei

Fueling and feedback

Françoise Combes
LERMA, Observatoire de Paris, CNRS, PSL University, Sorbonne University, F-75014 Paris, France
and
Collége de France, 11 place Marcelin Berthelot, F-75005 Paris, France

IOP Publishing, Bristol, UK

ISBN 978-0-7503-3035-0 (ebook)
ISBN 978-0-7503-3033-6 (print)
ISBN 978-0-7503-3036-7 (myPrint)
ISBN 978-0-7503-3034-3 (mobi)

DOI 10.1088/2514-3433/ac2a27

Version: 20211201

AAS–IOP Astronomy
ISSN 2514-3433 (online)
ISSN 2515-141X (print)

British Library Cataloguing-in-Publication Data: A catalogue record for this book is available from the British Library.

Published by IOP Publishing, wholly owned by The Institute of Physics, London

IOP Publishing, Temple Circus, Temple Way, Bristol, BS1 6HG, UK

US Office: IOP Publishing, Inc., 190 North Independence Mall West, Suite 601, Philadelphia, PA 19106, USA

Cover image: Centaurus A. Image credit: ESO/WFI (Optical); MPIfR/ESO/APEX/A.Weiss et al. (Submillimetre); NASA/CXC/CfA/R.Kraft et al. (X-ray).

Contents

Preface

Active galactic nuclei (AGN) are powered by supermassive black holes at galaxy centers. At the maximum of their power, they can radiate 1000 times more than their host galaxy. This activity must be fueled by gas infall, but the timescales to bring the gas have been a problem for a long time. Now with high resolution and sensitivity instruments, like the Atacama Large Millimeter/submillimeter Array (ALMA) in the radio, or MUSE at the European Southern Observatory (ESO)/Very Large Telescope (VLT), it is possible to unveil the interstellar medium and stars in the circumnuclear regions, discover molecular disks and tori, and the source of molecular outflows, the main feedback processes of AGN on their host.

The book contains a general review of all our knowledge of AGN, powered by super-massive black holes: how AGN are fueled by gas (or stars) and what are the problems encountered; how we can account for the wide variety of AGN, from Seyfert/Liners to quasars, and for the two types (AGN 1 and 2) either through dust obscuration, and geometrical orientation arguments, or through intrinsic activity difference, qualitative or quantitative. The fueling requires an exchange of angular momentum, which can be fostered by bars, spirals, and embedded features (bars within bars). After the duty cycle of the AGN has started, the luminosity is so high that the gas may be ejected (winds, radio jets), which stops the future star formation in the host-galaxy, and plays the role of feedback. Molecular outflows have been widely observed in the recent years, and numerous examples are described. Feedback can be triggered by winds when the activity of the AGN is near maximum, and through the radio jets, at low activity level. Different models have been devised to understand the structure of the circumnuclear regions, including a dusty or molecular torus, a bicone ejection, and thin or thick disk to collimate the flows. These models are now beginning to be better constrained by the high resolution radio observations, including masers mapped with very long baseline interferometers, or millimetric observations of tori. The components in the nuclear regions are frequently misaligned with the outer galaxy disk, which fosters the efficiency of AGN feedback.

Author Biography

Françoise Combes

Françoise Combes is professor at the College de France, on the chair "Galaxies and Cosmology." She is an astrophysicist at Paris Observatory, and member of the Academy of Sciences. Her main areas of research are the formation and evolution of galaxies, their dynamics and co-evolution with supermassive black holes, the interstellar medium of galaxies and dark matter models. By her computer simulations, she was the first to discover the mechanism to form bulges in spiral galaxies, through vertical bar resonances. She was also a pioneer in molecular absorptions in front of remote quasars, leading to constraints on fundamental constants variation.

Françoise Combes studied physics at the Ecole Normale Superieure (ENS), in Paris, and became ENS assistant professor in 1975. She was deputy director of the Physics Laboratory at the Ecole Normale Supérieure from 1985 to 1989, and became an astronomer at the Paris Observatory in 1989. She has received several awards, including the CNRS Gold Medal (2020), the Tycho Brahe Prize from the European Astronomical Society (2009) and the Gothenburg Lise Meitner Prize (2017). She was elected member of the European Academy (2009) and president of the French Astronomical Society (2002–2004). She has been an associate editor of *Astronomy & Astrophysics*, the European Journal, since 2003. She obtained in 2021 the international L'Oreal-Unesco award for Women in Science.

Chapter 1

Introduction

This first chapter introduces briefly our basic knowledge of active galactic nuclei (AGN), their nature, sizes, emission lines, their global spectrum. The central engine allows stable orbits, and their distance to the black hole depends strongly on the rotation level and relative angular momentum between the black hole and material. The wide variety of classes, and in particular the types 1 and 2, may converge to a universal AGN model, if the orientation and obscuration are taken into account. This unification paradigm has gained a lot of support from observations, although there remain intrinsic activity differences between all AGN types. Radio galaxies (with jets) represent about 10% of all AGN, and correspond to a low-luminosity phase of the nucleus, at low Eddington ratio. This introduces the two categories of AGN feedback, quasar mode, radiation-pressure driven, and radio mode, mechanically driven, that will be discussed in the following chapters.

1.1 Some Brief History

Nuclei of galaxies can pass through bright activity phases, which can outshine their whole galaxy hosts. This is however a relatively rare phenomenon, especially at low redshift, in the local volume. This phenomenon called active galactic nuclei (AGN) has been known for less than one century. These "peculiar galaxies" have been observed since the 1940s, well before their true nature was realized. They were discovered first at radio wavelengths.

Grote Reber (1940) discovered radio sources, in particular the well-known prototypical Cygnus A. He was the first radio-astronomer in the world to make a sky map at radio frequencies, even with no spatial resolution. He was following Karl Jansky (1933), engineer at Bell Telephone Laboratories, who was in charge in 1930 to determine the radio environment at 15 m wavelength, in order to identify the possible interferences for radio communications. Jansky built an antenna like a dipole of 30 m length, that could be rotated on tyres. He discovered a signal of unknown origin, and after several months of monitoring found it was maximum at

doi:10.1088/2514-3433/ac2a27ch1

the Milky Way, and in particular its center, Sagittarius. After his publication in 1933, there was no enthusiasm however for this radio emission of the "stars," and Grote Reber did not find a position in any observatory. He built his own radio-telescope in his backyard, a parabola of 9 m in diameter, looking at 1.8 m wavelength, and made radio sky maps, with a beam of 13°. He did not detect bright stars, nor planets, nor nebulae like Orion, but galaxy centers, the Milky Way and Andromeda at the beginning. The largest sources were in the constellation of Cygnus, Cassiopeiae, Canis Major, and Puppis, then Perseus.

Carl Seyfert (1943) discovered galaxies with bright nuclei, strong and very broad emission lines, in particular hydrogen lines of up to 8500 km s^{-1} wide. The width of the Balmer lines appears to increase with the absolute magnitude of the nucleus and with the ratio of the light in the nucleus to the total light of the galaxy. Seyfert was the first to classify several galaxies with quasi-stellar nuclei. Their nucleus is so bright in such a small size, that they show a diffraction pattern like stars in the photographic plates (cf Figure 1.1). Their flux varies like a power-law with frequency. According to the width of their hydrogen Balmer lines, they are now classified as Seyfert 1 or 2: Seyfert 1 with very broad lines and wings, and Seyfert 2 with narrower lines.

The AGN family is large, and one of the main physical classifications follows their luminosity. In Seyfert galaxies, the nucleus has at maximum about the same luminosity as the whole galaxy (typically a few $10^{10}L_\odot$ in the visible range). Quasars are more powerful, up to 100 or 1000 times the whole galaxy (typically a few $10^{13}L_\odot$). However, they are even rarer, and are therefore found at higher redshift. This is why they were discovered later, by Maarten Schmidt (1963). At their remote distance, they look like stars (quasi-stars), outshining their host so much that the latter is invisible.

In the years 1940–1950, radio observations had more and more spatial resolution and it was possible to identify optical counterparts to radio sources. The first high-resolution radio maps of the sky from the Cambridge interferometer show, beside galactic sources and radio galaxies, a class of objects looking like stars, but with a

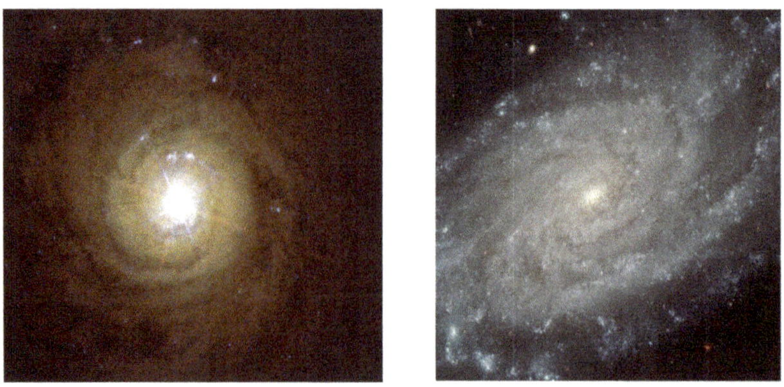

Figure 1.1. Left: NGC 5548, a Seyfert 1 galaxy. Note the diffraction pattern of the nucleus, which appears point-like, quasi-stellar. Right: NGC 3370, a normal galaxy, with a non-active nucleus, with no diffraction pattern, for comparison. Reprinted with permission from ESA/Hubble and NASA.

spectrum corresponding to nothing known (broad lines, continuum, absorption). They are called quasars, «quasi-stellar radio sources ».

In 1963, the Dutch astronomer Maarten Schmidt, who had emigrated to the United States in 1959, had the bright idea to identify the mysterious spectrum of the radio source 3C273 as a galaxy at $z = 0.158$, corresponding to a distance of 2 billion light years. This was a challenging idea, since this redshift ($z = \Delta\lambda/\lambda_{\rm rest} = \lambda/\lambda_{\rm rest} - 1$) was by far the highest at this epoch. It was 10 times higher than for most Seyfert galaxies. If these "stars" are such remote objects they are ultra-luminous. Some astronomers then questioned the cosmological redshift of these too luminous objects (Arp 1966, Hoyle & Burbidge 1966). Later, with the sensitivity and clean point spread function (PSF) of the Hubble Space Telescope (HST), it has been possible to subtract the central quasar source, and see the host galaxy behind, cf Figure 1.2, by Martel et al. (2003).

Seyfert galaxies have a strong UV continuum, and the flux increases with frequency as a power-law (cf Figure 1.3). In the optical, in a high state, very broad lines are observed with wings up to 10,000 km s^{-1} (cf Figure 1.4). The gas has a high-excitation state, and high levels of ionization (NeV, OVI). The nucleus emission is un-resolved at the HST resolutions (that is smaller than 10 pc). If the broad wings are due to virialized motions, huge masses can be derived in a small volume, up to $10^9\ M_\odot$ in 1 pc (applying the velocity $V^2 \sim GM/r$, in a radius r).

Two types of lines are observed in Seyfert galaxy spectra: permitted and forbidden lines. The permitted lines are the recombination lines, Lyman and Balmer lines of hydrogen, and also Mg II or C IV. Forbidden lines are called this because they do not appear in the laboratory. Their coefficient of spontaneous emission is so low that their corresponding upper states are de-excited by collisions,

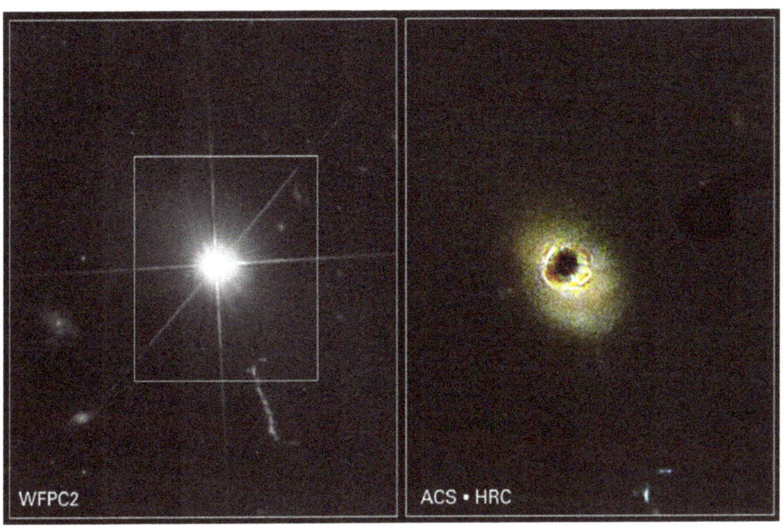

Figure 1.2. Left: optical image of the quasar and radio source 3C273, the first identified by Maarten Schmidt in 1963. Right: once the point-source subtracted, it is possible to see the light from the underlying host galaxy. Reprinted from Martel et al. (2003). © 2003. The American Astronomical Society. All rights reserved.

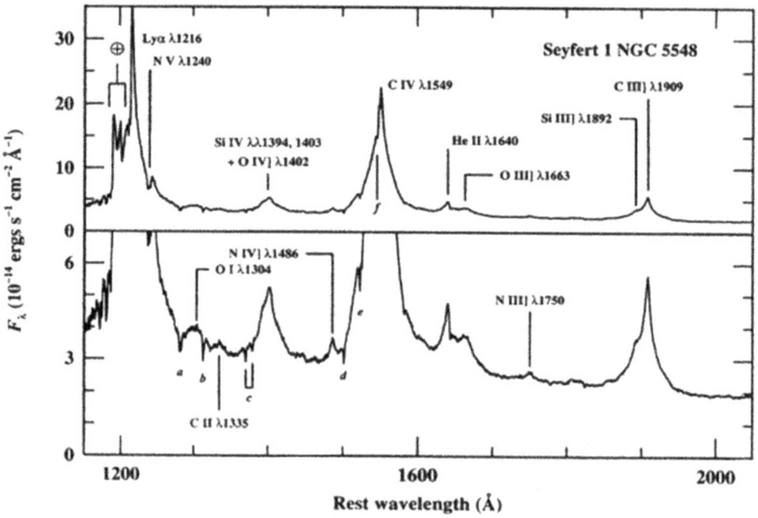

Figure 1.3. Nuclear UV spectrum of the Seyfert 1 NGC 5548. The panel at bottom is the vertically enlarged version of the top spectrum, to better show the broad wings. Reprinted from Korista et al. (1995).

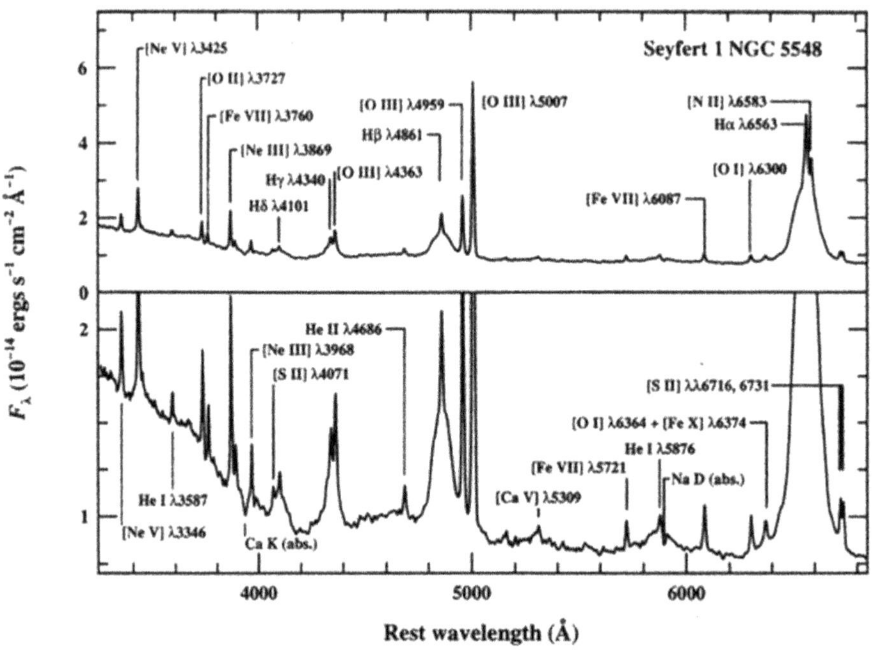

Figure 1.4. Optical spectrum of the Seyfert 1 NGC 5548 nucleus. The panel at bottom is the vertically expanded version of the top spectrum, to better show the weaker features. The full width at half maximum (FWHM) of the broad lines are 5900 km s^{-1}, and of the narrow components 400 km s^{-1}. The continuum rises at the blue end after 4000 Å in a "blue bump" made of Balmer lines and ionized iron FeII, here in a high state. Reprinted from Peterson et al. (1994).

between other ions or the walls in the laboratory. In interstellar space, the density is so low, that collisions are rare, and the spontaneous emission can be observed. However, there is a threshold density for that, meaning that in the high gas density near the nucleus, these lines will not be observed. That is why broad lines are only observed for permitted transitions. The forbidden transitions are noted in square brackets, such as the [O III] 4959, 5007 Å line. Some semi-forbidden lines are noted with only one square bracket, such as the [C III] 1909 Å. The latter are still broad lines, while the forbidden lines do not get broader than a few 100 km s^{-1}.

The broad line region (BLR) contains the gas close to the nucleus, rotating in an accretion disk. The width of the lines squared is proportional to the mass of the black hole in the center, and inversely to the distance from the center. In this region only the recombination lines are emitted, Lyα at 1216 Å, Hβ at 4861 Å, Hα at 6563 Å, and strong ultraviolet lines such as C IV at 1549 Å, MgII at 2800 Å. These permitted lines can be emitted at high density, with the electronic density n_e larger than 10^7 cm^{-3}, or even 10^9 cm^{-3}. Forbidden lines are emitted only in diffuse media, with density of the order of 10^3 cm^{-3}. These regions are much farther from the center, and the corresponding emission lines never have broad line wings. This region is called narrow line region (NLR). In these regions, typically observed are narrow lines of [O III] at 4959 and 5007 Å, [O II] at 3727 Å, [Ne V] at 3426 Å, [S III] at 9060 and 9532 Å. A large range of ionization may exist, from [O I] and [N I], up to [Ne V] and [O VI]. The latter are characteristic of an AGN (active nucleus), even the hotter stars cannot excite the gas at this level.

1.2 Excitation of Lines, Masses

The Balmer lines are expected to be emitted by an ionized gas recombining at a temperature ~10^4 K in the usual ratios Hα:Hβ:Hγ = 2.8:1.0:0.47, which are called decrements. The corresponding wavelengths of these lines are Hα 6563 Å, Hβ 4861 Å, and Hγ 4340 Å. If these ratios are not observed, but are higher, this means that the bluer lines (Hβ and Hγ) suffer from extinction by dust. The method, called the Balmer decrement method, is very useful to derive the dust column density. There could however exist gas at different temperatures, and with different excitation. Seyfert galaxies for example have a larger decrement than what is regularly found in HII regions photo-ionized by stars. Is this due to a different excitation, or to reddening due to dust? In the latter case, the continuum and the lines should have the same reddening. This is very useful to disentangle the physical phenomenon.

The relative strengths of the various emission lines can give important clues on the physical parameters of the ionized medium, temperature, density, and mass. The general features are common to ionized gas media in different astrophysical context, HII regions, planetary nebulae or supernova remnants. Using the photo-ionization equilibrium, we can derive a relation between the number dN/dt of ionizing photons emitted per second by the active nucleus, and the electronic volume density n_e, through the "ionization parameter" $U = dN/dt/(4\pi r^2 c n_e)$, translating that the ionization is compensated by recombination. This equilibrium occurs over a relatively narrow range of temperatures, 1–2 × 10^4 K. In contrast, the n_e densities can vary a lot, from larger than 10^9 cm^{-3} in the BLR to below 10^3 cm^{-3} in the NLR. Note that both in the NLR (300–500 km s^{-1}) or the BLR (up to 25,000 km s^{-1}), the motions of the gas is highly supersonic, since the sound velocity at 10^4 K is of the order of 10 km s^{-1}.

To compute masses from the emitted fluxes, we have to take into account that emissivities are proportional to the square of the density (n_e^2), being due to recombination. In other words, the flux of recombination lines is proportional to the product of the mass and density. Therefore, in the BLR at high density, only a small mass is sufficient, typically $M = 25M_\odot$. In contrast, in the larger and more diffuse NLR volume of forbidden lines, much more mass is necessary to explain the observations.

Narrow line Seyfert 1 (NLS1) galaxies are special cases (e.g., Osterbrock & Pogge 1985). They possess some characteristics of Sy1, as strong continuum, and highly excited lines, such as [Fe VII], [Fe X], however, they do not show broad permitted hydrogen lines, but only narrow permitted and forbidden lines. Sy2 have no broad lines either, but do not show high-excitation narrow forbidden lines.

1.3 Variability and Size

Most often, there is not enough spatial resolution in the observations to determine the size of the phenomena occurring around active nuclei. Fortunately, these phenomena are variable, over several timescales. The timescale is already an indication of the size where it is occurring. Imagine a periodic variation of the signal with a sinusoidal amplitude, with period T. Over a region of diameter d, the signal arriving to the observer from the remote side will arrive with a delay of d/c with respect with the signal from the front side, c being the light speed. And all the signals from elements in between will arrive with intermediate delays, or phase of the signal between 0 and $2\pi/T$ (d/c). All the variations will average out, if $d > cT$. Therefore, if the observer can detect variations on timescales of T, this means that the size of the varying region is smaller than cT. Some AGN are variable on timescales of days, or even hours, and the involved regions must be smaller than a light-day, or a light-hour (lt-h). This is the case for the continuum in ultraviolet and X-rays of certain quasars and Seyfert nuclei, and this means that their emission must come from the accretion disk close to the black hole. The emission lines vary on short timescales for the BLR, in general of the size of the solar system (6 lt-h to the Kuiper belt radius), while the NLR in contrast is much farther from the center, sometimes with a structure of radiation cones, and timescales of several years.

Radio jets (cf Figure 1.5) are often relativistic, and are even more variable. Electrons in the plasma are emitting synchrotron radiation, when spiraling in a magnetic field. The electrons can also lose their energy by Compton interaction with their emitted photons (mechanism SSC: synchrotron self-Compton). The radiation can be seen at multi-wavelengths from radio to X-rays. When the jet relativistic expansion is towards the observer, all timescales are shortened by relativity boosting effects. This is the Doppler beaming effect, responsible for flux enhancement and contraction of the variability timescales, Apparent motions are accelerated by the Lorentz factor $\gamma = (1 - v^2/c^2)^{-1}$. Quasars with axis oriented along the line of sight are called BL Lac objects or blazars. Their continuum emission is mainly due to the jet, and so much

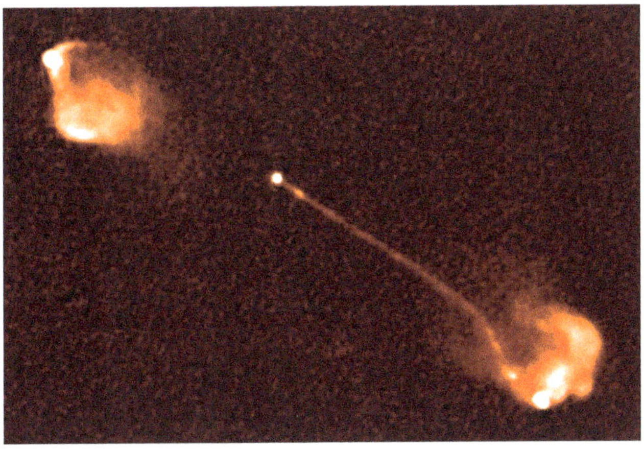

Figure 1.5. Radio image at 6 cm wavelength of the quasar 3C175, obtained with the VLA by Bridle et al. (1994). The redshift is 0.768. The vertical distance between the two most distant hot spots is 28", or 210 kpc. Reprinted from Bridle et al. (1994).

boosted that the lines are completely diluted and invisible; blazars are characterized by a featureless continuum emission, and strong variability on short timescales.

1.4 Global Spectrum SED

A typical quasar spectrum can be decomposed into several components, continuum and lines. The continuum consists of a power-law at low frequency, from the radio to infrared. Then in the mid and near-infrared (NIR) begins a blackbody emission from the accretion disk. At decreasing distance from the active nucleus, the temperature of the disk is increasing, and the total accretion disk emission can be decomposed in the outer (red), middle (green), and inner (blue) blackbody components with increasing temperature (cf Figure 1.6)

According to the AGN luminosity, the non-stellar radiation dominates more or less over the host stellar and interstellar radiation. This dominance is obtained when considering small apertures around the center. The various regions of the spectrum can be attributed to several physical mechanisms. At the low frequency end, the radio domain, the emission is a power-law due to synchrotron emission by relativistic electrons in a magnetic field. The slope of the power-law can give the age of the emission. Close to the AGN source, electrons have their full energy and the slope is small (almost flat spectrum). Then high-energy electrons lose their energy first, and the slope becomes steeper, unless there exist shocks to re-accelerate them. In some cases, the synchrotron can persist into the optical and X-ray domain. In the millimeter and sub-millimeter domains, the thermal emission reigns by dust heated by the AGN, in a molecular/dusty torus around the accretion disk. After that, at much harder frequencies, between 1 μm and 2 keV, we see the thermal emission from the accretion disk itself, which is optically thick (Figure 1.6). For the hard X-rays it cannot be thermal emission. It is either synchrotron emission, or inverse Compton, where softer radiation from the disk is reflected on the corona, and gives rise to

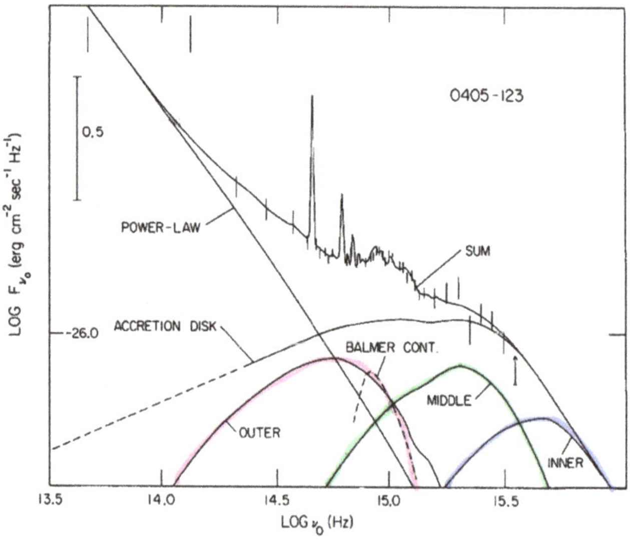

Figure 1.6. Decomposition of the quasar PKS 0405-123 ($z = 0.57$) spectrum in a power-law, and the accretion disk emission, with several blackbody components, at different temperatures. The inner disk is from 1.23 to 20 R_g, the middle disk ranges from 20 to 100 R_g, and the outer one until 1000 R_g, where R_g is the gravitational radius = GM_{BH}/c^2, or half the Schwarzschild radius of the central black hole. Reprinted from Malkan (1983).

harder X-rays, gaining energy from relativistic electrons. The corona is a hot plasma above the colder accretion disk, heated by reconnection of magnetic fields, generated by buoyancy instability in the cold accretion disk. At the high energy end of the spectrum, the γ-rays are also emitted by non-thermal processes, which range from inverse Compton in beamed relativistic jets to pair annihilation in the jets.

1.5 The Central Engine

Quasars are the brightest objects in the universe, thanks to gravitational energy released by matter falling into the central black hole. Seyfert galaxies are less luminous, but can still have up to 100 times the luminosity of the Milky Way. The Milky Way has a luminosity of $2 \times 10^{10}\ L_\odot \sim 8 \times 10^{36}$ W $= 8 \times 10^{43}$ ergs s^{-1}. This energy amounts in 1 Gyr to a total of $E = 2 \times 10^{60}$ ergs, and is of the same order of magnitude as its gravitational potential energy or kinetic energy, corresponding to its total mass ($M = 10^{12}\ M_\odot$), i.e., $GM^2/R \sim MV^2 \sim 8 \times 10^{59}$ ergs.

For a quasar, at least 10^{46} ergs s^{-1} is required. As soon as the existence of black holes was shown to be inevitable in general relativity (Penrose 1965), astrophysicists have built models based on supermassive black holes to provide this energy (Hoyle & Fowler 1963; Lynden-Bell 1969; Lynden-Bell & Rees 1971). Let us calculate the order of magnitude of this energy: for a mass m coming from infinity, the potential energy at the level of the horizon of the black hole of mass M, is GmM/R_s, with $R_s = 2GM/c^2$. R_s is the Schwarzschild radius, or horizon of the black hole, where the escape velocity becomes equal to the speed of light c. The available energy is therefore $\frac{1}{2}mc^2$, half of the mass energy. Observation shows that the efficiency of transforming this energy into radiation is such that the observer sees 10% of mc^2. The matter is heated by shocks, occurring when it spirals in. This energy is much more than the energy released into stars by nuclear fusion reactions, which results in less than 1% of mass energy. By consuming 1–10 solar masses per year, a black hole can radiate 1000 times more than the 200 billion stars contained in a galaxy like the Milky Way. But how much matter can the black hole swallow, and is there a limit to its greed?

When the flow of material swallowed is too great, the energy expended in radiation is such that the radiation pressure on the ionized gas compensates for gravity. The gas can no longer fall into the black hole, on the contrary it is pushed back, in a galactic wind, which can leave the galaxy. This limit corresponds to Eddington's maximum luminosity, $L_{\mathrm{Edd}} = 4\pi GcM\, m_{\mathrm{p}}/\sigma_{\mathrm{T}}$, where σ_{T} is the Thomson scattering cross-section for the electron and m_{p} is the mass of a proton. The Eddington luminosity is proportional to the mass M of the black hole, and equal to 3 10^{13} solar luminosities for a black hole of one billion solar masses. A quasar will not be able to radiate more than this limit, which corresponds to the accretion of 20 solar masses per year, for this black hole of one billion solar masses. The Eddington limit is obtained when considering the momentum of the radiation P_{rad} = energy flux $c = L/4\pi r^2 c$. The force due to the radiation on the ionized gas is then $P_{\mathrm{rad}}\, \sigma_{\mathrm{T}} = L\sigma_{\mathrm{T}}/4\pi r^2 c$ and equals the gravity force GMm_{p}/r^2 at the limit. A similar limit may be obtained farther from the central engine, on the dust component, and the cross-section is now σ_d, the equivalent dust cross section per proton, which is about 1000 times higher than σ_{T} (see also Section 3.2 of Chapter 3).

1.5.1 Last Stable Orbits

In a non-rotating black hole, matter can rotate in circular orbits around the center, but there does not exist stable orbits until the horizon. First there are unstable orbits, and then no circular orbits at all, as shown in Figure 1.7. The last stable orbit is called the ISCO (innermost stable circular orbit), and is located at 3 Rs.

However, the situation depends on the angular momentum J (or spin) of the black hole. In a rotating (Kerr) black hole, the Lense–Thirring effect, or frame-dragging effect, entrains all the matter to rotate in the same sense. This produces a gain of angular momentum, if the matter was previously rotating in the same sense (direct, or prograde), and therefore the matter can approach closer to the black hole, with a stable orbit. In the maximum rotating case, i.e., $a = c\ J/GM^2 = 1$, the direct stable orbit falls to 0.5 Rs. In contrast, when the gas is retrograde with respect to the black hole spin, the Lense–Thirring effect makes it lose angular momentum, and the stable orbit has a larger radius, up to 4.5 Rs. Some similar effect can be seen for massless particles, such as the photon, which can rotate in the circular photon ring, which is 1.5 Rs, for a non-rotating black hole (cf Figure 1.8).

The efficiency for the active nucleus to radiate energy depends a lot on the black hole spin. When $a = 0$, the gravitational energy released by the gas orbiting on the ISCO is only 6% of the total. When the gas is able to rotate in a stable manner closer to the black hole, for $a = 1$, the energy released on the ISCO is 30%. When the mass of the black hole is determined through the kinematics inside its sphere of influence, then through the variability of the matter on the accretion disk, it should be possible to derive the black hole spin.

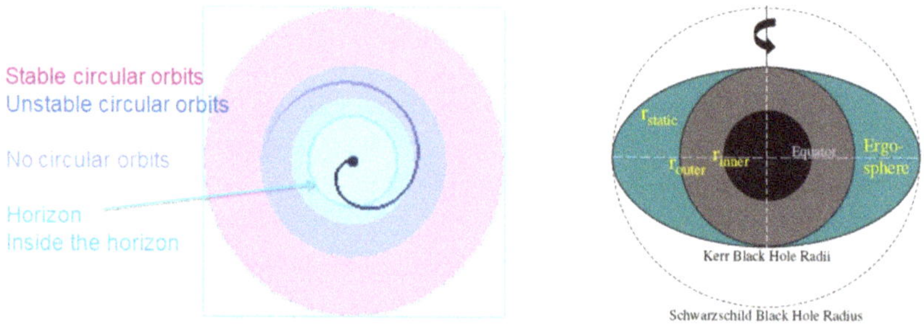

Figure 1.7. Left: stable and unstable orbits around a Schwarzschild (non-rotating) black hole. Right: a Kerr (rotating) black hole: definition of the ergosphere.

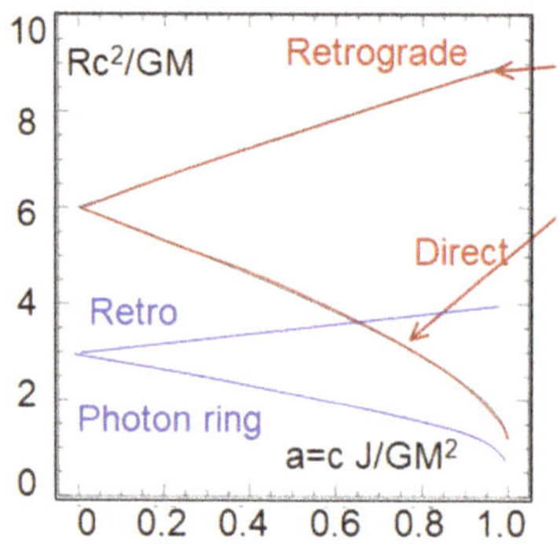

Figure 1.8. The last stable circular orbit around a black hole, as a function of its degree of rotation (angular momentum J), quantified by the dimensionless quantity a, comprised between 0 and 1. The two red curves correspond to gas rotating either in the same sense or retrograde with respect to the black hole spin, and the blue curves to the photon rings, in the equatorial plane.

In addition, the Lense–Thirring effect provides torques on the gas accreting with a tilted angular momentum, which is the most frequent case. The torques will warp the gas disk, and progressively force the gas to align on the equatorial plane: this is the effect first studied by Bardeen & Petterson (1975).

1.5.2 The Iron Kα Line

To find evidence of the black hole spin, we must find kinematic tracers of the matter very close to the black hole, on the accretion disk. The matter is almost fully ionized there and very hot, radiating in X-rays. The ionized iron (Fe) line exists in the X-ray domain. Iron is an element with atomic number 26, and Fe XXVI is equivalent to the hydrogen with only one electron.

When an X-ray photon collides with an Fe ion, removing an electron from an inner K or L shell, the ion may return to a lower energy state by emitting an electron from a higher shell (this is known as the "Auger effect")—or by a radiative transition. The relative probability of a radiative transition is the fluorescence yield. The energy of the Kα line depends on the number of electrons present and it increases as the line becomes more highly ionized, reaching 6.9 keV for Fe XXVI. The ionization state observed in active nuclei indicates gas temperatures which are relatively cool (about a million degrees) and the strength (i.e., the equivalent width) is quite high, indicating that the gas producing the Fe line has a high covering factor. This corresponds to the accretion disk, where X-rays are emitted just above in the corona.

The fluorescent Kα Fe line is the most powerful tool to determine the black hole spin, being redshifted and skewed by relativistic effects (cf Figure 1.9). The profile shapes are very different as a function of the dimensionless parameter *a*. It is not necessary to know the black hole mass in advance, and the inclination of the accretion disk is determined simultaneously with the measured spectrum. From the dozens of such spectra measured with precision, it has been revealed that a large fraction of supermassive black holes in AGN are rapidly-spinning. In recent years, the method has been made more robust, by considering all the spectrum (not only iron) and eliminating possible absorption (e.g., Reynolds 2014).

1.6 The AGN Zoo, Blazars, Quasars, and Others

AGN appear in various categories, which might show very different aspects, but are all of the same nature, originating from matter accretion onto a supermassive black hole. The various aspects could be due to the intensity (quasars are much more

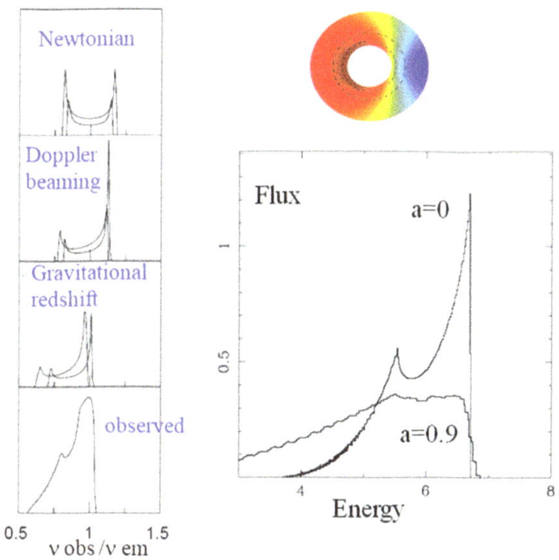

Figure 1.9. The Fe Kα line as a diagnostic for the black hole spin. The left column shows how the Fe line is predicted in the accretion disk, first with a normal Newtonian case, then with the relativistic effects: in particular, the frequency is globally redshifted, and asymmetric in favor of the blue-shifted side. The right panel displays the difference between the expected profiles for a non-rotating or highly rotating black hole. In the latter case, the profile is broadened, since the gas can rotate almost six times closer. Reprinted from Fabian et al. (2000). © 2000. The Astronomical Society of Pacific. All rights reserved.

powerful than Seyferts for instance), or due to projections effects (some Seyfert 2 having their broad line region obscured by dust, with respect to Seyfert 1), and others due to the phase in which we observe the AGN: the activity of nuclei are intermittent, and the presence of radio jets could also be intermittent. At the low-luminosity end, the activity of the nucleus may be confused with the starburst activity of the galaxy, especially at low spatial resolution. Therefore, it is sometimes difficult to disentangle several simultaneous activities, such as LINERS (low ionization nuclear emission line regions), composite objects, in transition between AGN and HII regions classifications.

When the accretion disk is projected face-on, and the jet is seen on the line of sight, the relativistic beaming enhances considerably the continuum emission, and the lines are then disappearing (cf Figure 1.10). These objects are called blazars, from the first of the series, the prototypic BL Lac. This configuration enhances the variability, this is in particular the case of optically violent variables (OVV), which vary on timescales smaller than a day.

Blazars are characterized by strong variability, within short timescales, down to a few hours. These could be due to cooling knots in the jet, or shocks and

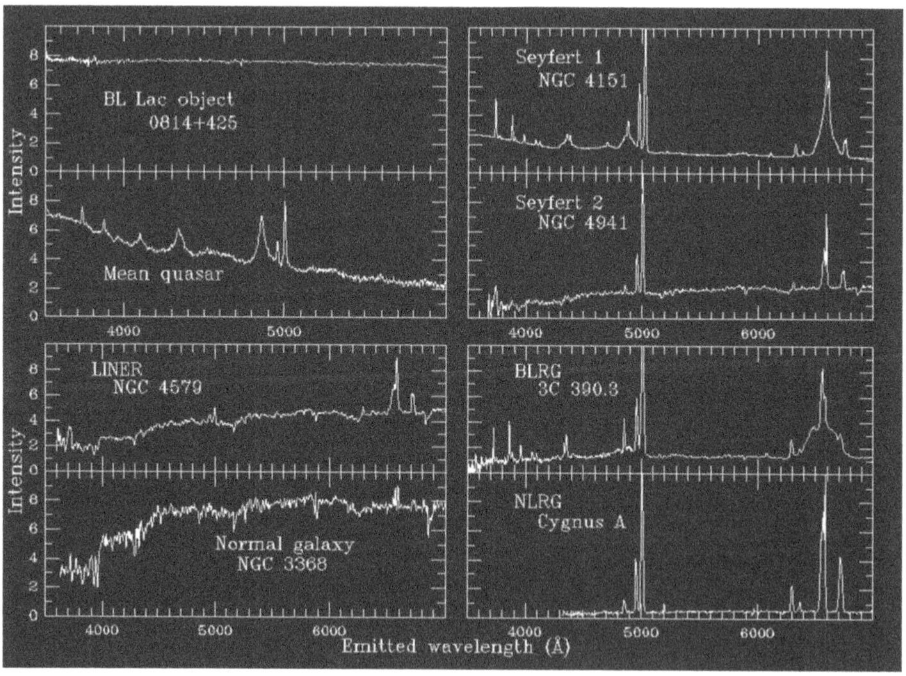

Figure 1.10. Comparison between several kinds of AGN, distinguished by their optical spectrum, between 3500 and 7000 Å (rest frame). At the top left, the blazar has a featureless continuum, with respect to the mean quasar just below. At the top right, the Sy1 and Sy2 galaxies are recognized by the presence or not of the BLR, as well as the broad line radio galaxies (BLRG) and their narrow line counterpart (NLRG) below. The bottom left displays a normal galaxy (non-active nucleus) spectrum for comparison, and a LINER galaxy, which is an intermediate or low-luminosity AGN. Courtesy of Bill Keel, from Bill Keel's page https://pages.astronomy.ua.edu/keel/agn/spectra.html.

turbulence, but also to geometrical features, such as deflection, change of the viewing angle along the line of sight. Comparison between the various wavelengths in the variability have shown that jet's re-orientation and twists, due to magnetic instabilities, could be the origin of longer (day) timescales, since they are achromatic. However, the intra-day variability depends on frequency, and should be intrinsic physical jet features. Following the relation between the size of the emitting region, and the timescale of observed variability, it has been established that these jet substructures occur on scales of 10^{-3} pc (e.g., Raiteri et al. 2021).

1.7 Emission Lines, and the BPT Diagnostics

In addition to studying the continuum at many wavelengths, which can bring much information on the few parsecs around the central engine, the various lines emitted in the extreme conditions of the accretion disk witness the physical conditions around the black hole. The large widths (1000–10,000 km s^{-1}) of the broad line region is a tool to measure the mass of the black hole. The various lines reveal the source of gas that feeds the central monster.

Star-forming galaxies also emit emission lines, however their spectrum is at lower frequency, limited by the temperatures of the young stars, while AGN can emit a much harder spectrum. To disentangle AGN from star formation in a galaxy, and know which radiation source is dominating the line excitation, it is current to build line diagnostics.

To quantify the degree of excitation and ionization of the medium emitting the optical lines, it is convenient to plot ratios between forbidden lines and Balmer recombination lines. Since the [N II] 6584 Å and Hα lines are very close in frequency, as well as [O III] 5007 Å and Hβ, these couples of lines will suffer from the same extinction, and their ratios will be exempt of these perturbations. The BPT diagrams (named after Baldwin, Phillips & Telervich) are precisely 2D diagrams with such line rations, the most famous being [N II]/Hα versus [O III]/Hβ (Baldwin et al. 1981). Such a diagram is displayed in Figure 1.11, and the various zones are delimited by curves, typically in a flying seagull shape.

Statistically, a large fraction of nearby galaxies reveals a low-luminosity activity of the nucleus, unrelated to stellar activity. Two thirds of early-type (ellipticals to early-spirals up to Sb) and 15% of late-type spirals, Sc to Sm, show a nuclear activity (Ho 2008). The others have a sleeping supermassive black hole. This increase of activity towards early-types is related to the observed proportionality between the mass of the bulge and mass of the black hole in a galaxy (e.g., Kormendy & Ho 2013).

The standard BPT diagram has been built to diagnose local galaxies. At high redshift ($z > 1.5$), galaxies on the main sequence were forming stars more actively, and their interstellar medium and corresponding lines had more extreme excitation, similar to starburst at low z. The classical BPT diagram has to vary to take into account this evolution, and to be able to separate AGN from star-forming galaxies, as shown in Figure 1.12.

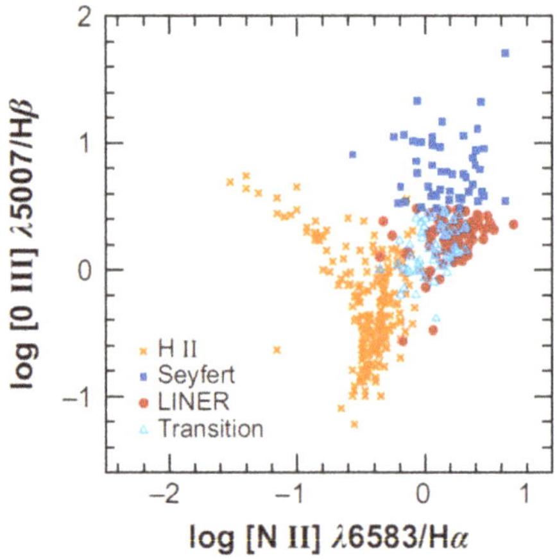

Figure 1.11. Standard BPT diagram, displaying by a symbol the location of each galaxy, dominated by star formation (orange), AGN (blue), LINER (red), and transition objects (empty triangles). Adapted from Ho et al (1997).

1.8 Unification Models

AGN are observed in a wide range of properties, not only in terms of luminosities, but also in terms of qualitative features, morphologies, spectra, and they reflect the wavelength range in which they have been discovered and selected. However, they are all powered by a central black hole, and they share common properties, so that there should exist interesting key parameters able to describe them all, making some order in the vast complex and confusing phenomenology of the AGN zoo. One of the key parameters is the projection effect, and the orientation of the line of sight with respect to the accretion disk. The realization that there might exist a dusty torus at sub-parsec scale obscuring the main engine, when seen edge-on, has helped to reduce the complexity of AGN classifications.

Around an active nucleus, there may exist two regions of optical emission lines, excited by the AGN: the NLR, and the BLR, according to their velocity width. The NLR are extended over kiloparsec in Seyfert-type galaxies, and their velocity range is lower than 1000 km s^{-1}. The density of the emitting medium is weak, between $n = 10^{3-6}$ cm^{-3}, lower than the critical density for the forbidden lines, which are present, in particular [O III]. The gas excitation reveals photo-ionized gas. The region may correspond to gas outflowing from the center, or only gas present in the line of sight of the AGN radiation, and illuminated by the nucleus. The range of velocity in the BLR reaches up to 40,000 km s^{-1}. The region is not resolved (much lower than 10 pc in size). The density is higher than the critical density for the forbidden lines, and only permitted lines are observed (densities $n = 10^{9-10}$ cm^{-3}). The temperature of the gas is of the order of 10,000 K from the observed ratio [C III] 977 Å/[C III] 1909 Å. The BLR is detected only in AGN of type 1 (Sy1 and quasars), while the NLR is present in both types 1 and 2 AGN.

The presence of narrow low and high ionization in forbidden lines (such as [O III], [Ne III], [O II], [O I], [N II], [S II]), and several very high ionization coronal lines (such

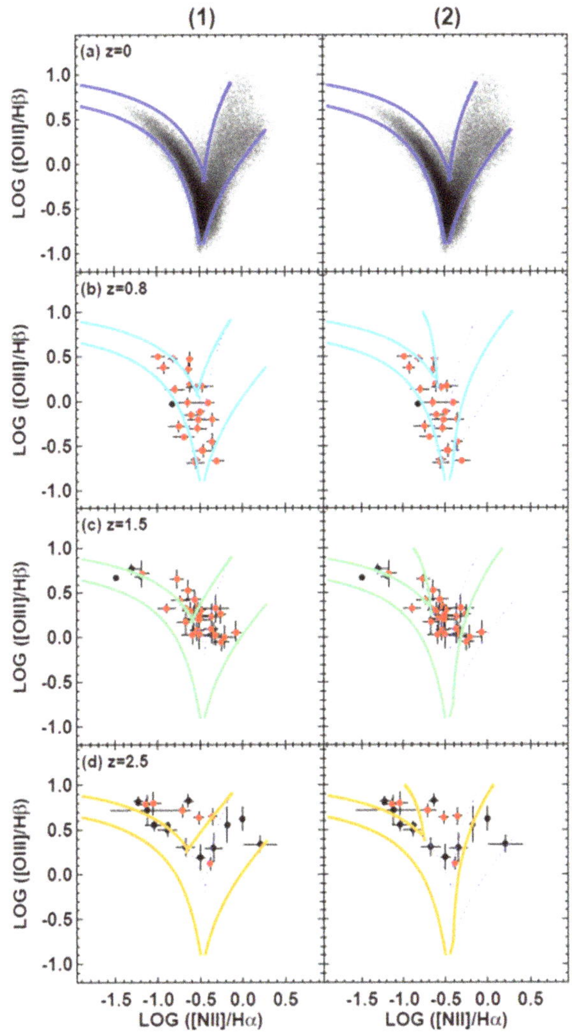

Figure 1.12. Evolution of the BPT diagnostics with redshift z: in each diagram, the dotted blue lines repeat the $z = 0$ lines for comparison. The left column is the solar metallicity objects (1), and the right column the low metallicity ones (2). Reprinted from Kewley et al. (2013). © 2013. The American Astronomical Society. All rights reserved.

as [Fe x], [Si x]) in both Sy1 and Sy2 with similar line ratios, suggests that both Sy types come from the same objects. In addition, it has been possible to detect the BLR lines in some Seyfert 2 (for instance in the Sy2 NGC 1068 or Mrk348; Antonucci & Miller 1985; Miller & Goodrich 1990) in polarized light, reflected on the electrons of the ionized region. This means that the photons exiting from the accretion disk are blocked by obscuration on the line of sight towards the observers, but can come back towards them, after reflection. Of course the polarized light is much smaller than the original one, and this is possible to detect only in the stronger nearby sources. The size of the obscuration should be larger than the accretion disk region, while smaller than the NLR region, and it should cover all azimuths, like a torus aligned with the accretion disk. Indeed, if it was only one cloud along the line of sight, reflected light could then come from all directions, and the resulting polarization would be zero.

Confirmation of the obscuration of the BLR in Sy2 was obtained by the detection of the broad lines in the NIR, such as Paα, Paβ or Brγ, in galaxies where they were not seen in the optical, since dust absorption can be several times higher than in the NIR (Veilleux et al. 1997; Lutz et al. 2002). Similar evidence is found in the X-ray bands, when soft X-rays (energy lower than 2 keV) are absorbed in Sy2, while hard X-rays (>2 keV) are detected, suggesting column densities along the line of sight of at least $N_H = 10^{22}$ cm^{-2} (Bassani et al. 1999). When the column density becomes larger than 10^{24} cm^{-2}, only very hard X-rays 10–100 keV can get out, and the nucleus is said to be Compton-thick. In this regime, it might still be possible to detect the Fe Kα line at 6.4 keV, either highly absorbed, or scattered by the circumnuclear medium. Since the continuum is deeply depressed by absorption, the equivalent width (EW, which is the ratio of line to continuum) is enhanced, and an EW higher than 500eV is a sign of a Compton-thick nucleus (Matt et al. 1999).

If this assumption of a dusty torus is correct to explain the absorption of the BLR continuum and lines in type 2 AGN, then the shape of the NLR should have a bi-conical morphology, since the photons from the accretion disk are collimated by the torus (see Figure 1.13). This ionization cone is indeed observed in many objects (e.g., Barbosa et al. 2009). The NLR region has typical sizes of 10–100 pc, and can reveal only ionized gas in rotation, or also contain ionized gas outflows. The opening angle of the ionization cone is wide (half angle of ~45°), and the axis of the cone might not be aligned with the minor axis of the large-scale galaxy disk. It is better aligned with the radio jet, when there exists one, although it might have slight angles offset. There is evidence of interaction between the two, the radio jet producing shocks in the circumnuclear medium, favoring [O III] emission (Capetti et al. 1996).

The dusty and molecular torus is assumed to be composed of dense clouds, able to subsist to the strong tidal shear in the vicinity of the black hole. The torus size is assumed to be of pc size, which corresponds to the 2–4 pc size of the mid-infrared dust emission, obtained with the VLT interferometer (Jaffe et al. 2004). This means that it is under the sphere of influence of the black hole, in a Keplerian potential. The torus is also assumed to be geometrically thick, to provide the required obscuration. However, the mechanism to keep it thick, in the presence of dissipation (cloud–cloud collisions and radiation) is not well known: radiation pressure or stellar feedback might not be sufficient. It could be that the molecular torus is unstable through warping and bending processes, which thicken its appearance. Alternatively, it is likely that at this distance from the AGN, hydromagnetic and radiative processes could drive a wind from the disk, and the clouds just extracted close to the disk might be optically thick enough to provide the toroidal obscuration (Elitzur & Shlosman 2006). This outflow in addition would help to remove the angular momentum of the accreting gas. Support from this view has been brought by the

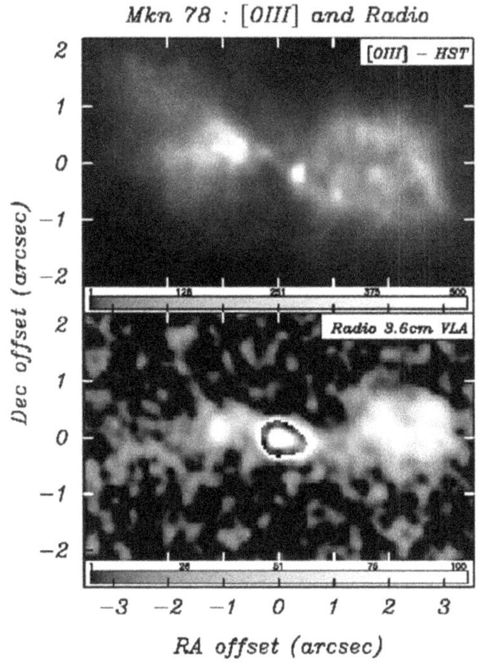

Figure 1.13. Emission line and radio images of Mrk78 ionization cone. Top is the [O III] image from HST; bottom is the VLA 3.6 cm radio image, from Whittle & Wilson (2004). To show the core more clearly, the center has been rescaled by a factor 5 division, above 200 μJy per beam. Reprinted from Whittle & Wilson (2004).

observations of water masers: either in a warped and very thin disk (NGC4258, NGC 1068; Greenhill et al. 1995; Gallimore et al. 1996), or also in a wind above the disk, with proper motions (e.g., NGC 3079; Kondratko et al. 2005).

The most recent observations of the dust distribution in the center of nearby Seyferts with the VLT interferometer in mid-infrared revealed that the dust is mainly in a polar extended structure, in addition to a thin disk (e.g., Hönig et al. 2013; Asmus 2019). This may originate from a dusty wind forming a hollow cone in the narrow line region (Hönig 2019). The wind will begin at the sublimation radius, which corresponds to the temperature above which grains do not survive. This radius of course depends on the luminosity of the nucleus, but might be of the order of 0.1 pc, and will be the start of the molecular dusty torus. From this radius, the disk wind could form the hollow polar cone, participating to the strong obscuration at its base. A schematic view of all components is drawn in Figure 1.14.

The clumpy nature of the torus, and even of the dust-free clouds inside the sublimation radius, is confirmed by the ubiquitous variability of X-ray absorbing columns densities in Sy2 galaxies (Risaliti et al. 2002). The variability timescales range from months to several years, and the amplitude from 20% to 80%. Some variabilities of the order of days, require velocity of the clouds larger than 1000 km s^{-1}, and are clearly on the BLR. In the case of NGC1365 for instance, the variability has

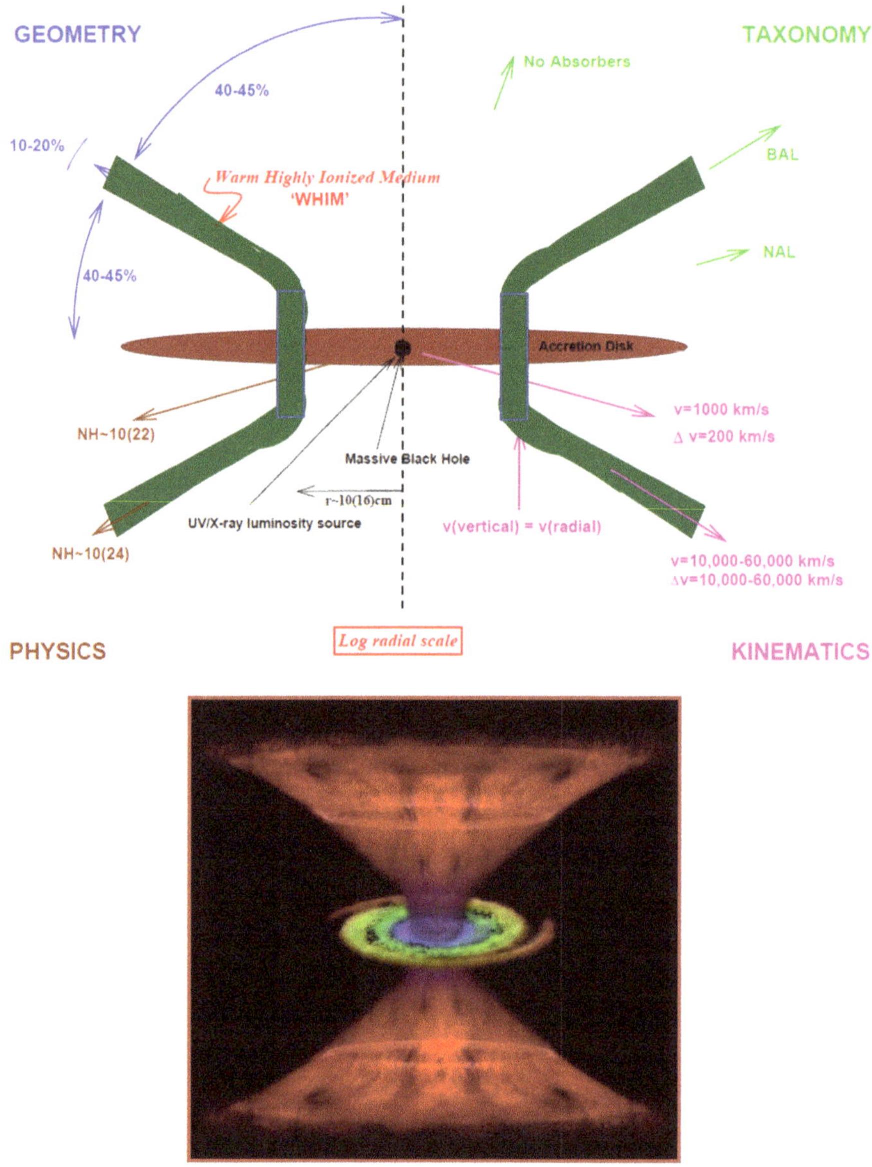

Figure 1.14. Possible structure of the AGN circumnuclear medium. The four quadrants indicate successively the geometry sketch, with the opening angles of the polar cone; the taxonomy, or spectroscopic appearance to a distant observer at various angles; the kinematics, with outflow velocities along different lines of sight; and the physics, with different characteristic sizes, and some typical column densities. BAL and NAL mean, respectively, broad and narrow absorption lines. The bottom picture is a 3D rendering of the structure, with colors indicative of temperatures. Reprinted from Elvis (2000). © 2000. The American Astronomical Society. All rights reserved.

revealed a cometary shape for the BLR clouds, likely due to shocks (Maiolino et al. 2010). This particular shape helps to explain the covering factors of clouds. Those clouds have a few months life-time, and are certainly replenished from accretion. The clumpiness in the dusty torus is supported by the large range of temperatures observed in the spectral energy distributions (SED), requiring very different expositions, and a range of densities (Nenkova et al. 2008).

The unification paradigm based on orientation also explains nicely the BL Lac category, where the continuum is so strong, that no emission line appears (cf Figure 1.10). The relativistic boost of the continuum emission towards the observer makes it dominant. In spite of the large success of the obscuration and orientation of the line of sight to explain the dichotomy between types 1 and 2 AGN, it is not the whole story. There exists nevertheless "true" Sy2, i.e., not only obscured Sy1, but objects that have actually and intrinsically no BLR. The most convincing phenomenon, showing that obscuration is not the only factor, is the changing look AGN. Some galaxies can change in timescales of decades from Sy1 to Sy2 and vice-versa (e.g., Mrk1018, McElroy et al. 2016). According to multi-wavelength data, and in particular in X-rays, this change cannot be attributed to varying obscuration, but must be attributed to a varying accretion rate. More generally, the existence or the width of the BLR can be directly related to the accretion rate, and therefore the luminosity of the AGN. The BLR width of Sy1 vary over a wide range, from ~1000 km s^{-1} in narrow-line Seyfert 1 (NLS1), to about 20,000 km s^{-1} for the most powerful ones. As a class, Sy1 are more luminous than Sy2. At low luminosities, the BLR of Sy2, normally still visible in polarized light or in the NIR, disappears all together. Many proposed models converge to predict the disappearance of BLR at low luminosity. The large widths of BLR lines are due to both the high rotational velocities of the material very close to the black hole, and also to the high-velocities of the outflowing wind. The wind is triggered as soon as the accretion rate, and the corresponding luminosity is close to the Eddington rate: then the radiation pressure becomes comparable to the thermal pressure of the gas. The wind is then outflowing vertically from the disk, at the radius of equalized pressures. The radius where the permitted BLR lines form depends both on the total luminosity and mainly on the X-ray spectrum, which can ionize the gas much farther from the center. With a steep X-ray spectrum, the broad emission lines, formed at a larger distance, have a smaller width, from the Kepler potential. Models then expect that the accretion rate is inversely proportional to the broad lines width, and this is what is observed for the NLS1, who have the highest relative accretion rate (e.g., Nicastro, 2000). The AGN with the largest (20,000 km s^{-1}) line widths have amongst the largest black hole masses, and low accretion rates. Larger widths should not exist, which is compatible with observations. Below a certain accretion rate, and luminosity, the BLR disappears, and this can explain the existence of unobscured Sy2.

At the opposite extreme, there exists some AGN where the NLR is so obscured that it is not detected in optical lines. This may occur when the nucleus is almost isotropically obscured and buried, and only some NIR or X-ray is piercing out (Imanishi et al. 2010).

1.9 Radio Galaxies, and Radio Properties of AGN

About 10% of AGN are strong radio sources, they were called radio loud (RL), compared to the others: radio quiet (RQ). However, this definition depends strongly

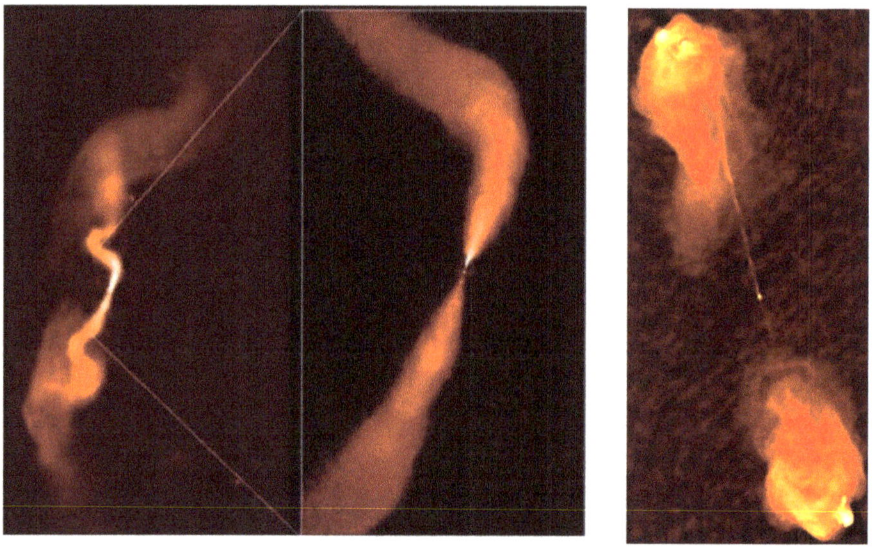

Figure 1.15. The two main categories of radio galaxies, according to Fanaroff & Riley (1974). The source on the left is the FRI source 3C31, which is brightened at the center, and the two jets are fading with distance. The right source is the FRII source Cygnus A, which is brightened at the ends. Both images have been obtained at VLA (NRAO). Credit: NRAO/AUI/NSF (CC BY 3.0): https://public.nrao.edu/gallery/quicklook-samples/.

on the sensitivity, and many more are now detected in radio. A better definition is now to characterize jetted and non-jetted sources (Padovani 2017). Radio emission can come from the AGN, and consist of a core emission and two relativistic jets, whatever their size. Even without nuclear emission, the disk of the galaxy can show some synchrotron emission, due to star formation and supernovae remnants, that will not be considered here.

Fanaroff & Riley (1974) classified radio galaxies into types I and II based on the ratio of the separation of the highest surface brightness regions on opposite sides of the central galaxy, and the extent of the source measured from the lowest surface brightness contour (see Figure 1.15). The ratio is either lower than 0.5 (FRI), or higher (FRII). This means that FRI are core brightened, while FRII are edge-brightened. The separation corresponds also to a luminosity classes, the FRII being globally more powerful, and rarer. Another class, FR0, has been introduced by Baldi et al. (2015), which is even more core-dominated than FRI.

As shown by the Figure 1.16 and the distributions of Eddington ratios, the radio galaxies are the weakest AGN, relatively accreting and radiating very inefficiently.

1.9.1 Efficiency of AGN

A nucleus is called active when its supermassive black hole accretes at an Eddington ratio $L_{\rm bol}/L_{\rm Edd}$ at least 10^{-5}, where $L_{\rm bol}$ is the bolometric luminosity of the nucleus. Our Galaxy, the Milky Way, is not active according to this definition. When the Eddington ratio is high, larger than 0.01, the accretion rate is sufficient to build a geometrically thin, but optically thick disk, and since the cooling rate is proportional to the square of density, this accretion disk is radiatively efficient. As can be seen in

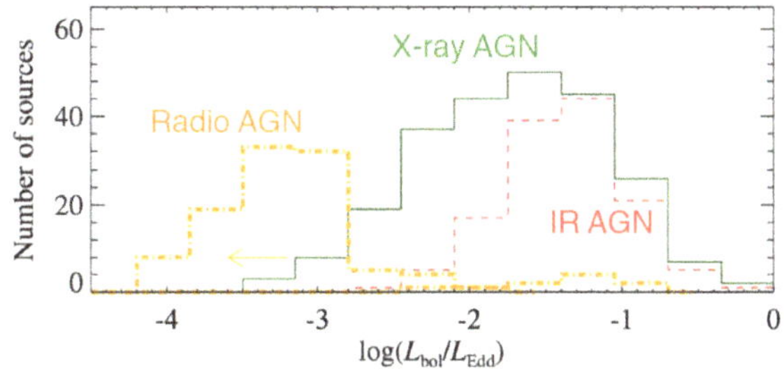

Figure 1.16. The distribution of the Eddington ratio L_{bol}/L_{Edd} for radio, IR, and X-ray selected AGN. Reprinted from Hickox et al. (2009). © 2009. The American Astronomical Society. All rights reserved.

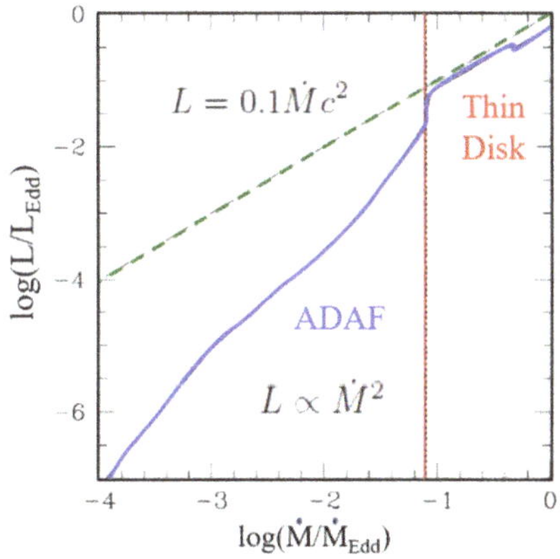

Figure 1.17. The bolometric luminosity vs. mass accretion rate, both normalized to Eddington, according to the model developed by Esin et al. (1997). The vertical red line corresponds to the critical accretion rate. Above this critical value, the accretion is via a geometrically thin disk, optically thick (or slim disk). Below this value, the accretion is via an advection dominated accretion flow, first at small radii only, and then over the whole disk. The luminosity is then much less, since the viscously generated energy is advected into the black hole. Reprinted from Esin et al. (1997). © 1997. The American Astronomical Society. All rights reserved.

Figure 1.17, the luminosity of the AGN is maximum, proportional to the accretion rate, with an average conversion ratio of mass into energy of about 10%. This corresponds to the high state of Figure 1.18. At the very high end of this category, when the Eddington ratio is near 1, radiation pressure can trigger a wind from the disk, since it is sufficient to compensate gravity.

When the accretion rate is much lower than the critical rate of 0.01 Eddington, then the accretion regime becomes inefficient. Several denominations characterize this regime, radiatively inefficient accretion flow (RIAF), or advection dominated accretion flow (ADAF), since most of the energy is advected into the black hole without being efficiently radiated. Instead, the gas unable to cool efficiency heats and inflates, as shown in Figure 1.18. This low state is often accompanied by a relativistic radio jet, which ejects most of the energy of the AGN through kinetic energy. This low state today is encountered mostly in very massive black holes, at the center of early-type galaxies, for which only a very low gas density is available in their

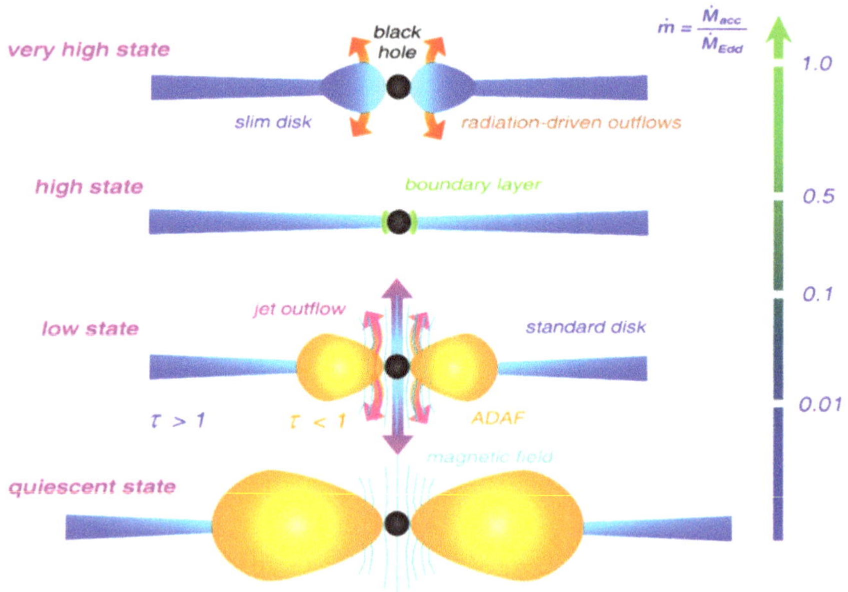

Figure 1.18. The various steps of the accretion flows, according to the Eddington-normalized accretion rate, indicated in the right vertical line. At the bottom, the density of the flow is so low that its cooling is not efficient: the flow heats and become a thick cloud. At higher accretion rate, a slim cold disk begins to form in the outer part, while the center is an ADAF, with radio jet. At higher accretion rate, there is an ideal cold slim disk, and the accretion disk is efficiently radiating. Above, the luminosity is so high, that the radiation pressure triggers a disk wind. Reprinted with permission adapted from Esin et al. (1997), redrawn by A. Muller (2004, private communication).

circumgalactic medium. The high state, at the opposite is frequently encountered in young objects, with still a less massive black hole, in a late-type galaxy, rich in gas. This picture appears to be true at higher redshift also, and the co-evolution of galaxy bulges and black holes is likely the fact of secular evolution more than mergers (Kormendy & Ho 2013; Heckman & Best 2014).

In addition, it has been observed that the X-ray spectrum emitted by the AGN is softer in the high state, and harder in the low state. This is remarkably similar between AGN/supermassive black holes, and X-ray binaries, where the mass of the black hole is stellar (e.g., Fender & Belloni 2004). This is explained if the accretion disk is surrounded by a hot corona of ionized gas, where the soft X-ray photons from the disk gets energy from the electrons of the corona through Comptonization (inverse Compton scattering) into the hard X-ray regime. The corona is present essentially in the low state, and its role might also be played by the relativistic jet itself, present essentially in the low state.

1.9.2 High and Low Excitation for Radio Sources

As already mentioned earlier, see Figures 1.16 and 1.18, radio sources are among the less efficient AGN, although they are among the most massive galaxies. However, it is still possible to distinguish two categories of radio sources the high-excitation ones (HERG) and the low excitation radio galaxies (LERG), as detailed by Best &

Heckman (2012). The first ones have an Eddington ratio between 0.01 and 0.1, while the LERG have an Eddington ratio below 0.01. They are on both sides of the critical accretion rate, which distinguish the radiation efficiency. Although it is possible to find examples of both categories at all radio luminosities, there is a tendency for HERG to have higher radio luminosity as well. As for the feedback mechanism, and the triggering of gas outflow, the HERG would be in the quasar mode, where a wind is ejected from the disk by radiation pressure, while LERG prefer the radio mode, where gas is shocked and entrained in the radio jets.

These two excitation classes are based on the optical line diagnostics, essentially on the [O III]/Hβ ratio, and BPT-like diagrams. If all radio sources are located in the AGN region, some of the weaker ones are in the Sy/LINER zone. To be more precise and separate more clearly the classes, an excitation index, combining the various ratios ([N II]/Hα, [S II]/Hα, [O I]/Hα, etc) has been considered (Buttiglione et al. 2010). This excitation index separates the two classes very well (cf Figure 1.19). Another separating parameter is the broad line of permitted lines (Hα, Hβ), which exists mainly in HERG. However, the difference between the two classes is not only a quantitative effect, since there exist powerful LERG, and the hypothesis has been made of a qualitative difference, where accretion in LERG is powered by hot gas (they are mainly massive early-type galaxies), and in HERG by cold gas, allowing a slim accretion disk and broad lines.

1.9.3 Radio Jet Formation

Jets are composed of a plasma of both electron/proton pairs and electron/positron pairs. Both are required to reproduce the X-ray radiation spectrum. Radio jets are collimated from pc-scales to Mpc-scales, and this wide range of scales poses a challenge for the interpretation. The collimation through only thermal pressure is insufficient, and strong magnetic fields are essential (Sauty et al. 2002).

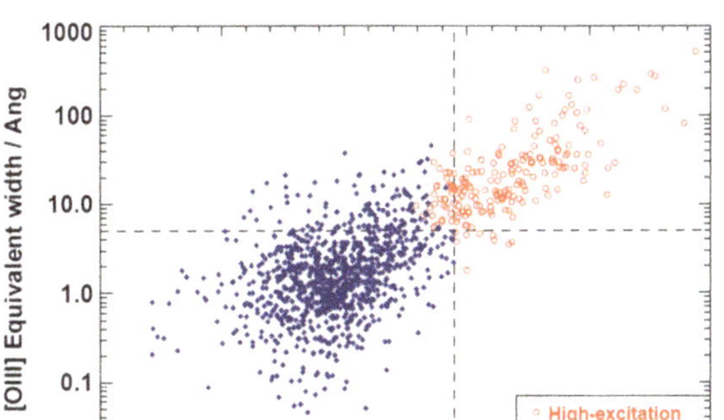

Figure 1.19. The [O III] line equivalent width (in Angstrom), versus the excitation index, based on optical lines (Buttiglione et al. 2010). The two radio galaxies classes, HERG and LERG, are then very well separated. Reproduced from Best & Heckman (2012). © 2012. The Royal Astronomical Society. All rights reserved.

The power of the radio jet depends in a large part on the black hole spin, since jets are produced by extracting energy from the rotational kinetic energy of the black hole. In the Blandford-Znajek (1977) mechanism, the rotational energy of the accreting black hole is extracted by a large-scale magnetic field through a pure electromagnetic process. The vacuum surrounding the Kerr black hole is then unstable, the electromagnetic energy creating relativistic electron/positron pairs. The accretion disk itself plays the major role in the Blandford-Payne (1982) mechanism, where the angular momentum is removed magnetically from the cold disk, through a hot corona above the disk. The field lines stressed by rotation are collimating the wind in a jet. The collimation might be helped by an encircling MHD wind leaving the accretion disk at a slower speed. Powerful FRII radio sources are formed when the hole spins rapidly and the relativistic core accelerates the MHD sheath. It has been proposed that the spine of the jet is composed preferentially by electron/positron pairs, while the electron/proton plasma is around (Sol & Pelletier 1989). The Faraday rotation measure has shown that the magnetic field is rotating helicoidally in the jets, and there might exist an electric current in the spine, coming back in the sheath, similar to a coaxial cable (Contopoulos et al. 2009).

Bibliography

Antonucci, R. R. J., & Miller, J. S. 1985, ApJ, 297, 621
Arp, H. 1966, Sci., 151, 1214
Asmus, D. 2019, MNRAS, 489, 2177
Baldi, R. D., Capetti, A., & Giovannini, G. 2015, A&A, 576, A38
Baldwin, J. A., Phillips, M. M., & Terlevich, R. 1981, PASP, 93, 5
Barbosa, F. K. B., Storchi-Bergmann, T., Cid Fernandes, R., et al. 2009, MNRAS, 396, 2
Bardeen, J. M., & Petterson, J. A. 1975, ApJ, 195, L65
Bassani, L., Dadina, M., Maiolino, R., et al. 1999, ApJS, 121, 473
Best, P. N., & Heckman, T. M. 2012, MNRAS, 421, 1569
Blandford, R. D., & Znajek, R. L. 1977, MNRAS, 179, 433
Blandford, R. D., & Payne, D. G. 1982, MNRAS, 199, 883
Bridle, A. H., Hough, D. H., Lonsdale, C. J., et al. 1994, AJ, 108, 766
Buttiglione, S., Capetti, A., Celotti, A., et al. 2010, A&A, 509, A6
Capetti, A., Axon, D. J., Macchetto, F., et al. 1996, ApJ, 469, 554
Contopoulos, I., Christodoulou, D. M., Kazanas, D., & Gabuzda, D. C. 2009, ApJ, 702, L148
Elitzur, M., & Shlosman, I. 2006, ApJ, 648, L101
Elvis, M. 2000, ApJ, 545, 63
Esin, A. A., McClintock, J. E., & Narayan, R. 1997, ApJ, 489, 865
Fabian, A. C., Iwasawa, K., Reynolds, C. S., & Young, A. J. 2000, PASP, 112, 1145
Fanaroff, B. L., & Riley, J. M. 1974, MNRAS, 167, 31
Fender, R., & Belloni, T. 2004, ARA&A, 42, 317
Gallimore, J. F., Baum, S. A., O'Dea, C. P., et al. 1996, ApJ, 462, 740
Greenhill, L. J., Henkel, C., Becker, R., et al. 1995, A&A, 304, 21
Heckman, T., & Best, P. 2014, ARA&A, 52, 589
Hickox, R. C., Jones, C., Forman, W. R., et al. 2009, ApJ, 696, 891
Ho, L. C., Filippenko, A. V., & Sargent, W. L. W. 1997, ApJS, 112, 315
Ho, L. 2008, ARA&A, 46, 475
Hönig, S. F. 2019, ApJ, 883, 171

Hönig, S. F., Kishimoto, M., Tristram, K. R. W., et al. 2013, ApJ, 771, 87
Hoyle, F., & Fowler, W. A. 1963, MNRAS, 125, 169
Hoyle, F., & Burbidge, G. R. 1966, ApJ, 144, 534
Imanishi, M., Maiolino, R., & Nakagawa, T. 2010, ApJ, 709, 801
Jaffe, W., Meisenheimer, K., Röttgering, H. J. A., et al. 2004, Natur, 429, 47
Jansky, K. G. 1933, Natur, 132, 66
Kewley, L. J., Maier, C., Yabe, K., et al. 2013, ApJ, 774, L10
Kondratko, P. T., Greenhill, L. J., & Moran, J. M. 2005, ApJ, 618, 618
Korista, K. T., Alloin, D., Barr, P., et al. 1995, ApJS, 97, 285
Kormendy, J., & Ho, L. 2013, ARA&A, 51, 511
Lutz, D., Maiolino, R., Moorwood, A. F. M., et al. 2002, A&A, 396, 439
Lynden-Bell, D. 1969, Natur, 223, 690
Lynden-Bell, D., & Rees, M. 1971, MNRAS, 152, 4
Maiolino, R., Risaliti, G., Salvati, M., et al. 2010, A&A, 517, A47
Malkan, M. A. 1983, ApJ, 268, 582
Martel, A. R., Ford, H. C., Tran, H. D., et al. 2003, AJ, 125, 2964
Matt, G., Guainazzi, M., Maiolino, R., et al. 1999, A&A, 341, L38
McElroy, R. E., Husemann, B., Croom, S. M., et al. 2016, A&A, 593, L8
Miller, J. S., & Goodrich, R. W. 1990, ApJ, 355, 456
Nenkova, M., Sirocky, M., Ivezic, Z., & Elitzur, M. 2008, ApJ, 685, 147
Nicastro, F. 2000, ApJ, 530, L65
Osterbrock, D. E., & Pogge, R. W. 1985, ApJ, 297, 166
Padovani, P. 2017, NatAs, 1, 0194
Penrose, R. 1965, PhRvL, 14, 57
Peterson, B. M., Berlind, P., Bertram, R., et al. 1994, ApJ, 425, 622
Raiteri, C. M., Villata, M., Carosati, D., et al. 2021, MNRAS, 501, 1100
Reber, G. 1940, ApJ, 91, 621
Reynolds, C. S. 2014, SSRv, 183, 277
Risaliti, G., Elvis, M., & Nicastro, F. 2002, ApJ, 571, 234
Sauty, C., Tsinganos, K., & Trussoni, E. 2002, LNP, 589, 41
Schmidt, M. 1963, Natur, 197, 1040
Seyfert, C. 1943, ApJ, 97, 28
Sol, H., & Pelletier, G. 1989, MNRAS, 237, 411
Veilleux, S., Goodrich, R. W., & Hill, G. J. 1997, ApJ, 477, 631
Whittle, M., & Wilson, A. S. 2004, AJ, 127, 606

Chapter 2

AGN Feeding

The fueling of the central engine is confronted with a huge angular momentum problem, even if the gas reservoir around the black hole is abundant. Low-mass black holes may accrete the stars in their close vicinity, especially with low angular momentum (almost radial orbits). Then it takes a long relaxation time to refill the loss-cone. Gas can cool and dissipate energy, and may be driven in through gravity torques from the non-axisymmetric potential, due to bars and spirals. Embedded nuclear bars and spirals. Primary bars can exchange angular momentum with the gas inside corotation, which will spiral inwards until the inner Lindblad resonance. Then an embedded nuclear bar can take over. When gas is reaching the sphere of influence of the black hole, the torque turns to negative, fueling the center. Growing a black hole from a stellar-mass seed is long, and the observation of very massive black holes in the early universe is a puzzle that could be solved through direct collapse of massive clouds into black holes.

2.1 Black Hole Growth in Symbiosis with Their Host Galaxy

There exist famous scaling laws between the mass of the central black holes (BH), and the mass of their host galaxies, suggesting that they both grow in symbiosis. Black hole masses are in general difficult to obtain, except in the case of our nearby one in the center of the Milky Way, due to its proximity (Genzel et al. 2010), or due to exceptional spatial resolution, like in the case of M87 in Virgo, where the Event Horizon Telescope Collaboration et al. (2019) has imaged the shadow. For the galaxies of our local environment up to our super-cluster Laniakea, the distance is of the order of 10–80 Mpc. The gravity of the black hole has only a very small sphere of influence (SoI), defined as the region where it dominates the gravity of the stellar bulge, i.e., its radius is $R_{\mathrm{SoI}} = GM_{\mathrm{BH}}/\sigma_{\mathrm{v}}^2$ = 10–50 pc depending on the mass of the galaxy. Here M_{BH} is the black hole mass, and σ_{v} the central stellar velocity dispersion. With the angular resolution (0.1″) of the Hubble Space Telescope (HST) it has been possible since the 1990s to measure the speeds of stars and gas

doi:10.1088/2514-3433/ac2a27ch2

very close to the black hole of most nearby galaxies. At a distance of 20 Mpc, 0.1″ equals 10 pc, therefore we can solve the black hole's sphere of influence for the most nearby galaxies, or farther for the more massive. At about the same epoch, adaptive optics on ground-based telescopes have achieved comparable resolutions. When the speed $V(R)$ of stars or gas are known at the radii R, the black hole mass can be estimated, exactly proportional to the product V^2R, since the potential is Keplerian. In the radio domain, even higher spatial resolution, of the order of a fraction of milli-arcsec, can be achieved with the Very Long Base Interferometry (VLBI). This technique is possible, when there exists maser emission (H_2O, OH, NH_3) in a thin disk around the black hole. The monitoring of the maser-cloud motion can also give high precision distances (Greenhill et al. 1995).

One of the first relationships, linking the mass of the black hole and that of the bulge, was established by Magorrian et al. (1998) with 36 galaxies observed with the HST. The relation was established with the bulge luminosity, easy to convert to mass, since the mass-to-light ratio of bulges does not vary much from galaxy to galaxy. The relation is about linear, and reveals that the black hole mass is only 0.1% to 0.2% of the bulge mass. Very soon after, it was realized that the relation is even less scattered when $M_{\rm BH}$ is plotted versus the central velocity dispersion, and it became the $M{-}\sigma$ relation. An example of this relation is shown in Figure 2.1. The relation is then no longer linear, but a power law, with an exponent 3–4. This is easy to understand, since there exists an empirical scaling relation between luminosities and velocity dispersions, with the same exponent 3–4, for ellipticals and bulges (Faber & Jackson 1976; Falcon-Barroso et al. 2011). It is not yet established why the $M_{\rm BH}$ relation is tighter with the dispersion σ than with the bulge mass or luminosity. It might be intrinsic or due to the fact that the dispersion and kinematics around the black hole are used to derive its mass.

The $M{-}\sigma$ scaling relation suggests that black holes are growing at the same pace as the stellar spheroids in galaxies. Exceptions to the rule are teaching us about the galaxy evolution. There are exceptions for instance for a number of early-type objects, where central black holes are too light for their velocity dispersion. The vast majority of these objects are interacting galaxies that will merge into one. The velocity dispersion has already increased in the interaction, but the black holes have not yet merged (Kormendy & Ho 2013). When the coalescence ends, the final system will be back on the relation, after less than a billion years.

There is also a whole category of objects, where the black hole is overweight, compared to the host galaxy. These galaxies are in dense environments, and they are often the brightest galaxy in the cluster (BCG). Their positions at the center of the potential well favors multiple galaxy mergers; they have swallowed many small companions, and therefore they are called cannibal galaxies. Galaxies in clusters have lost their cold gas, through tidal interactions, ram-pressure stripping and lack of cold gas replenishment, since the intracluster medium (ICM) is very hot (million degrees). The galaxy mergers in the center are therefore dry mergers, without star formation. In the BCG, there is frequently gas cooling from the hot ICM, with very low star formation efficiency. The central black holes are then swallowing gas before it has even had time to form stars. This process could make the black hole grow

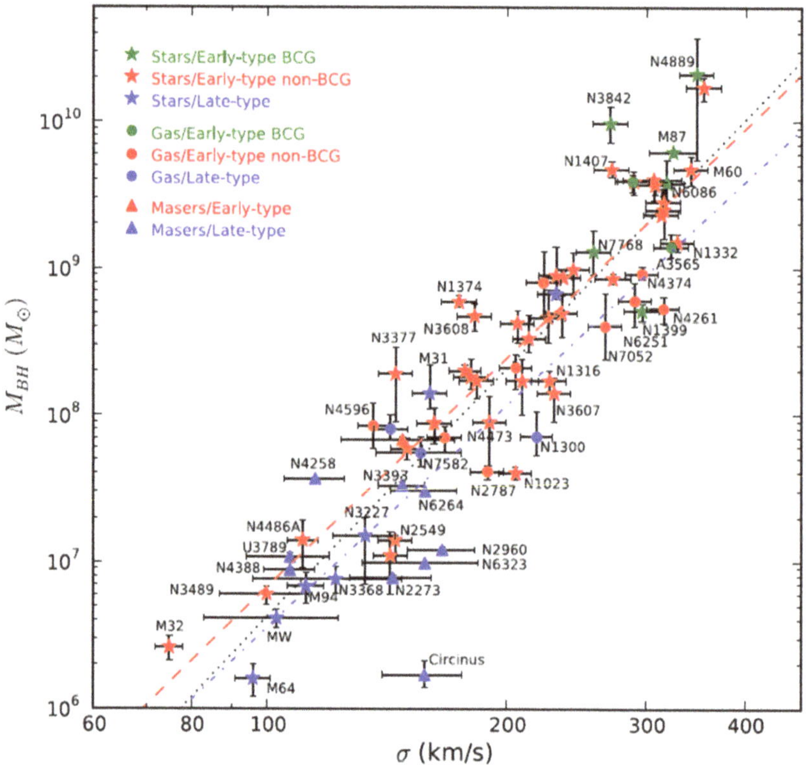

Figure 2.1. The M–σ relation, with 72 galaxies. The brightest cluster galaxies (BCGs) are plotted in green, other elliptical and S0 galaxies are plotted in red, and late-type spiral galaxies are plotted in blue. The black hole masses are measured using the dynamics of masers (triangles), stars (stars), or gas (circles). The different color power law fits correspond to the galaxies of the same color. Reprinted from McConnell & Ma (2013) © 2013. The American Astronomical Society. All rights reserved.

faster than the stellar bulge. The BCGs (green points) in Figure 2.1 are above the M $-\sigma$ relation. One example of over massive black holes is the galaxy NGC 1277 in the Perseus cluster. It is a lenticular galaxy, possessing a rotating stellar disk, and a massive bulge. Van den Bosch et al. (2012) measured a supermassive black hole mass of 1.7×10^{10} $M_\odot$, or more than half the mass of the bulge, instead of the standard ratio of 0.2%. This large value has been questioned by Emsellem (2013), who reduced by a factor of 4 the mass of the black hole estimated by stellar kinematics, and proposing a higher bulge mass, with a possible perturbed velocity field due to a bar on the line of sight. The mass of the black hole has however been confirmed with the kinematics of molecular gas (Scharwächter et al. 2016).

Other examples are found in the Perseus cluster, where NGC 1277 is the prototype, and in the center of other galaxy clusters. The stellar populations of these galaxies are very old. They appear as "relic" galaxies where all the stars have been formed at the beginning of the universe, in the first 1 Gyr. Across the next ~13 Gyr, they could have acquired gas to feed their black hole, but without forming new stars (Trujillo et al. 2014; Ferré-Mateu et al. 2017). Their center is compact and red, so they are called red nuggets. These relic galaxies might have swallowed satellites, but during their interaction the companion is tidally disrupted, due to the compact primary. Most of the companion stars are accreted at the periphery of the central massive core. This explains why these red quenched massive galaxies increase their size with time, at nearly the same mass and central dispersion (Newman et al. 2014).

There is also an important parameter related to the over massive black holes: the presence of a core or not in the central stellar distribution. Elliptical galaxies are known to be separated in two classes: core and coreless galaxies. This parameter is tightly linked to the morphology and kinematics, and also to the mass. Massive ellipticals have a boxy-shape spheroid, and are dominated by anisotropic velocity dispersion, while less massive ones have a disky shape, and more rotation (e.g., Cappellari 2016). The massive ones are rarer and have a core, while the others are coreless. The core is thought to arise in dry major mergers, while the coreless galaxies may be formed in gas-rich mergers, where the gas driven to the center by dissipation contributes to a peaked stellar distribution, creating a surplus of light through star formation. During a merger of two early-type galaxies each possessing a supermassive black hole, the dynamic friction causes the two galaxies to spiral towards each other, but also the black holes, which orbit fairly quickly in the center of the merger-remnant elliptical galaxy. Before merging the pair of black holes must lose their angular momentum by dynamic friction on the stars of the spheroid: these stars gain energy, and are evacuated from the center, which creates a deficiency of light in the center, and a core (Hopkins et al. 2009). In the M–σ relation, the over massive black holes happen to be in the core galaxies, and precisely in the relic objects, preferentially found in galaxy clusters. It appears that the relation becomes vertical in the diagram (cf. Figure 2.1), while the velocity dispersion saturates.

Other parameters have been studied in order to test their influence on the black hole mass. It has been shown that $M_{\rm BH}$ does not correlate with pseudo-bulges, those formed in secular evolution, while they correlate with classical bulges, thought to be formed in minor and major merges (e.g., Kormendy & Ho 2013). Barred galaxies have lower $M_{\rm BH}$ with respect to unbarred ones (Graham et al. 2011). Disks and dark matter are not particularly related to the black hole mass, beyond their correlation to the total mass of the galaxy, a global size effect.

Although the main correlation of black hole mass is with the bulge mass, there still exists rare bulgeless galaxies, essentially dwarf galaxies, which show AGN activity, and some have $M_{\rm BH}$ measured by reverberation mapping, from their broad Balmer lines (e.g., Dong et al. 2012). They have in general high accretion rates, and their $M_{\rm BH}$ cover the intermediate mass range of 100–10^5 $M_\odot$. It is now established that they follow the M–σ relation, although with much scatter (Greene et al. 2020).

2.2 How to Feed a Supermassive Black Hole

To feed the central black hole, the matter around the nucleus must lose a lot of angular momentum. Since the latter is proportional, per unit mass, to the circular velocity and to the radius, it is always increasing with radius, as shown in Figure 2.2.

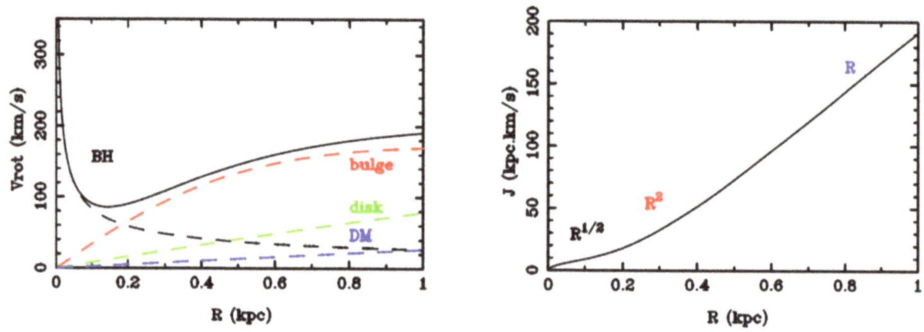

Figure 2.2. The angular momentum problem to feed the black hole: at left is displayed a typical rotation curve in a spiral galaxy, with a central zone dominated by the BH (Keplerian potential), then a rising velocity, dominated by the bulge, and finally a flat rotation (disk and dark matter DM). The various contributions of the mass components are marked in color; at right, the corresponding angular momentum per unit mass, which increases as power-laws of the radius, with slopes first ½ (Keplerian), than 2 and then 1.

For the material in the vicinity of the center (~1 kpc), it will be necessary to arrive on the accretion disc, whose radius is typically 10^{-2} pc, i.e., to lose a factor of 10^3 in specific angular momentum. The rate at which the mass has to inflow (*dM*/*dt*), to account for the AGN luminosity, assuming a radiative efficiency of 10%, meaning $L = 0.1 dM/dtc^2$, is varying from 10^{-3} $M_\odot$/yr to 2 $M_\odot$/yr for the most luminous quasars of 10^{13} $L_\odot$. Assuming a typical uptime of the order of 100 million years, the black hole would have to swallow about 2×10^8 solar masses during its duty cycle, and this gas has to come in just a few dynamical timescales, since the rotation period at 1 kpc radius is 30 Myr. Very efficient dynamic mechanisms are required to exchange the angular momentum at the kpc scale in the galaxy.

At these kpc-scales, the cold gas density is small, clouds are experiencing only little collisions per rotation, and the viscosity torques are negligible (e.g., Lin & Pringle 1987). In order to bring in the gas in such a short time, and cause it to lose its angular momentum, gravity torques are necessary. This means tangential forces, produced by non-axisymmetric disturbances, such as bars and spirals. If the disk remains axisymmetric, the gravitational forces are only radial, and there are no torques. Stellar bars create a robust and long-lived potential, that is able to drive the gas inwards. But some feeding of the black hole can also be due to the stars in the vicinity of the nucleus. This source is certainly important at the beginning of the universe, when stars of all orbits are present. After a first phase, where the low angular momentum stars have been swallowed, stellar relaxation takes a long time to replenish them.

2.2.1 The Swallowing of Stars

The contribution of stars in the BH feeding is enhanced by the existence of a nuclear star cluster, and is relatively more important for dwarf galaxies (e.g., Pfister et al. 2020). When a star travels close to the black hole, it can be destroyed by tidal forces, reduced to gas, and then swallowed by the black hole's accretion disk. Part of the material will be ejected at high speed, but the majority will be captured by the hole, causing a bright flare lasting at most a few years (Rees 1988). This luminous event is called tidal disruption event (TDE). To have a TDE, it is necessary that the mass of the black hole is lower than the Hills limit, of $3 \times 10^8\ M_\odot$. Indeed, a star begins to be tidally distorted when the average density of the perturber (the black hole) inside a sphere of radius corresponding to the relative distance star–BH becomes equal to the internal stellar density. The average density $<\rho>$ of a black hole inside its horizon (of radius $R_s = 2GM_{BH}/c^2$) is a decreasing function of its mass: $<\rho> \propto M_{BH}/R_s^3 \sim M_{BH}^{-2}$. This density becomes equal to the average density inside a star of the main sequence, for $M_{BH} = 10^8\ M_\odot$. For more massive black holes, the stars are no longer destroyed by the tidal effect outside the horizon, but inside, and no longer contribute to AGN activity (Hills 1975). Although the capture event is not visible, the mass of the black hole grows nevertheless. For dwarf galaxies, these bright TDE flares are one of the best way to discover low-luminosity dormant intermediate mass black holes (e.g., Greene et al. 2020). Their frequency has been estimated from observations to be of the order of 10^{-4}/yr per galaxy (van Velzen 2018; Auchettl et al. 2018). Most of TDE events are detected during surveys searching for supernovae (for example All-Sky Automated Survey for Supernovae: ASASSN). They are recognized by their peculiar light curve, composed at the beginning by a bright, super-Eddington accretion phase, followed by a $t^{-5/3}$ decline with time t (Rees 1988; cf Figure 2.3). More than 40 TDE have been identified through X-ray, optical and UV flares; they are preferentially in quiescent low-mass galaxies, which show post-starburst features (French et al. 2020).

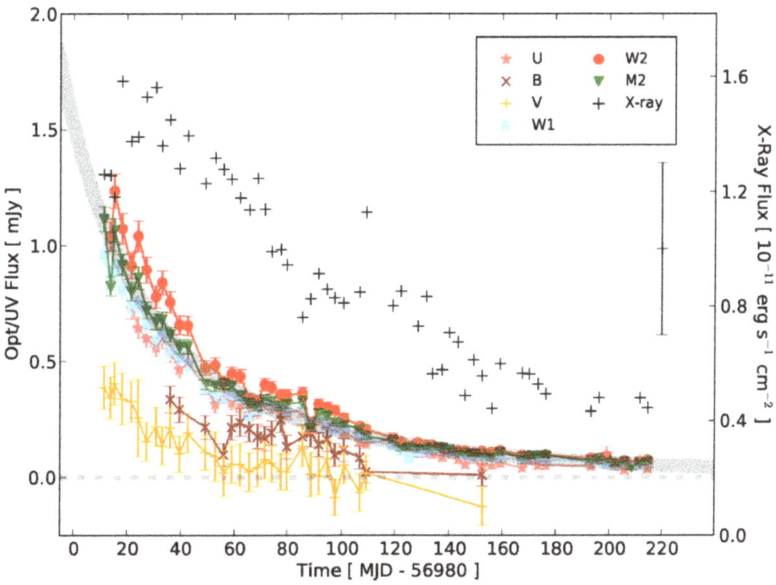

Figure 2.3. An example of tidal disruption event (TDE) detected at several wavelengths: X-ray, optical (U, B, V) and UV (W1, W2, M2) light curves of ASASSN-14li. The galaxy, at z=0.0206 is one of the closest TDE found. Reprinted from Miller et al. (2015), with permission from Macmillan Publishers Ltd.

For the BH to accrete and swallow stars, the latter have to arrive with low angular momentum, so that the pericenter of their orbit is close to the horizon. The last stable orbit depends on the spin of the black hole, and also on the prograde and retrograde character of the orbits, as was shown in the first chapter. At pericenter, the velocity of the material is close to c, and it can be shown that the critical angular momentum to be captured by a non-rotating black hole is $4GM_{BH}/c$. These values have to be changed for a rotating black hole, plus or minus depending on the sense of rotation, and for a rotating BH it also depends on the inclination of the orbit with respect to the spin axis. To be captured, a star must therefore arrive in what is called a loss-cone (e.g., Merritt 2013). In its first growth phase, the black hole swallows all the stars in the loss-cone, but when it is emptied, the process is stopped, and two-body relaxation is required between the stars to repopulate the low-cone. The 2-body relaxation timescale can be shorter than the Hubble time for small and compact systems, like globular clusters, or nuclear star clusters. For massive bulges and ellipticals, this is not true, and the loss-cone effect will stop rapidly the accretion, unless there are gravity torques, due to a strong triaxiality, or some other perturbation (galaxy interaction, merger). Another possible mechanism is the interaction between the stars and the accretion disk around the black hole. At each crossing the stars lose energy, which circularizes their orbit. They become closer to the center, and could be captured faster (MacLeod & Lin 2020).

The presence of a nuclear star cluster can help the black hole growth, through another process, collisions between stars. In the dense stellar cluster existing very close to the nucleus, the velocities of stars as they approach the black hole are relativistic, and collisions can tear off a lot of mass, or even destroy stars (Rauch 1999). Collisions also have the advantage of redistributing angular momentum, thus refilling the loss-cone, repopulating the low angular momentum stars, which can then be swallowed by the black hole. Collisions between stars are now a favored way to form black holes of intermediate mass in dense star clusters (Reinoso et al. 2018; Sakurai et al. 2019).

There is also competition for gas accretion between star formation and black hole growth. During a galaxy merger, a large quantity of gas is driven inwards by the gravity torques. This gas can feed both a nuclear starburst and the supermassive black hole (Norman & Scoville 1988). Even after the gas infall episode, stars from the cluster can continue to grow the black hole. Stars lose mass over their lifetime, on average for a star cluster, at a rate of 10^{-11} $M_\odot$/yr per $M_\odot$ of star. If there is a nuclear star cluster in the few central parsecs, then it is possible to have enough gas that has already lost its angular momentum to form the star cluster. A gas-rich galaxy merger is able to form a cluster of 4×10^9 $M_\odot$ and therefore providing up to ~10^7 $M_\odot$ in one cycle of AGN activity of 100 Myr, almost simultaneously with the starburst. It should also be taken into account that stars in the vicinity of a black hole are submitted to a tidal stress, and may have lost their envelope, or be swollen, therefore more fragile and sensitive to mass loss. It appears that nuclear star clusters might be formed by the spiraling in of globular clusters through dynamical friction in dwarf galaxies (M^* lower than 10^9 $M_\odot$), while they are formed *in situ* by gas *accretion* in a merger in more massive galaxies (Neumayer et al. 2020).

According to simulations of an ensemble of stars near a black hole, the equilibrium density distribution should show stellar cusps, meaning a density rising as a power law towards the center, where it is singular. The cusp is established through stellar dynamics, and can take several slopes, according to initial conditions, and formation scenarios ending with a slope near 2 as an asymptotic radial distribution for a relaxed star cluster (Alexander 2017). Observations of the stellar cluster in the Milky Way, and nearby galaxies have indeed revealed a stellar cusp (e.g., Habibi et al. 2019; Rouan et al. 2019). This high density favors the interaction with the black hole and its growth.

2.2.2 The Physics of Gas Accretion

Galaxy disks at kpc scales are unstable to non-axisymmetric features, spirals and bars. These instabilities are constantly renewed due to gas dissipation: gas cloud collisions trigger shocks and radiation, where the kinetic energy is dissipated. At the opposite, the stellar component is collisionless. This is due to the large number of stars ($N \sim 10^{11}$) in a galaxy disk. Stars are so numerous that they create a smooth potential, which does not scatter their orbits. In other words, the timescale of 2-body relaxation, T_{rel}, when stellar encounters can deviate significantly their trajectory, is much larger than the Hubble time. More precisely the ratio of T_{rel} to the crossing time $tc = R/V$ (R and V being the characteristic size of the system, and V the velocity of the stars) is $T_{rel}/t_c \sim h/R\ N/(8 \log N)$, h being the characteristic height of the disk. For typical disks, t_c is of the order of 10^8 yr, and $T_{rel}/t_c \sim 10^8$. The 2-body relaxation has to be considered only for small systems, such as globular clusters, or nuclear stellar clusters.

The stability of disks is described at small scale by the Jeans criterion, i.e., the velocity dispersion for the stars, the turbulence for the gas, stabilizes the small scales, smaller than the critical $\lambda_J = \sigma/(G\rho)^{1/2}$, where σ is the velocity dispersion, ρ the density. At large scales, the rotation stabilizes, through the tidal shear, all scales larger than $L_{crit} = G\Sigma/\kappa^2$, where Σ is the surface density and κ the epicyclic frequency. The disk stability is ensured for a dispersion σ such that $\lambda_J = L_{crit}$. This defines a critical dispersion $\sigma_{crit} \sim \pi G\Sigma/\kappa$, so that the stability criterion is $Q = \sigma/\ \sigma_{crit} > 1$ (Toomre 1964). Q is the Toomre parameter.

When the disk is unstable and forms spiral features, these instabilities heat the disk, developing the velocity dispersion and streaming motions, which stabilize the stars (Toomre parameter $Q >> 1$). But the gas is dissipative and radiates this energy. The Toomre parameter returns back to 1 for gas, which can be subject to other instabilities, and so continuously. These instabilities, constantly renewed, transfer mass and angular momentum, and are at the origin of an effective kinematic viscosity (Lin & Pringle 1987).

Spiral and bars are density waves (Lin & Shu 1964). They constitute an immaterial pattern, in which the stars and gas clouds move and rotate at a different speed. To explain the robustness of bars, and the low-winding of spirals, they must have a constant angular pattern speed with radius. This pattern speed of the wave is called Ωp. The motion of matter particles in the galaxy disk can be described at the zeroth order as a rotation around the center at the angular velocity Ω, and at first order by radial and azimuthal oscillations about this rotation, with the epicyclic frequency κ. The disk gravitational potentials are such that the frequency $\Omega-\kappa/2$ is only slowly varying with radius, and if a pattern has Ωp equal to this value, the self-gravity of particles trapped in the wave will make it constant (e.g., Toomre 1981). In the referential frame of the pattern, rotating at Ωp, this means that the orbits of particles are almost closed. They trace two epicycles, while making one turn around the galaxy. Therefore, their orbit is elongated, looking like an ellipse. The frequency $\Omega-\kappa/2$ is the precessing rate of these elliptical orbits.

When there is a rational ratio between the epicyclic frequency κ and the angular speed ($\Omega-\Omega p$) in the pattern rotating frame, the orbits are closed, and this corresponds to a resonance between the motions of the bar and the disk matter. After Bertil Lindblad, who discovered these in the 1930–1940s, we can distinguish three types of resonances, the inner Lindblad resonance (ILR), the corotation (CR) and the outer Lindblad resonance (OLR). These are characterized by $\Omega p = \Omega-\kappa/2$, $\Omega p = \Omega$ and $\Omega p = \Omega + \kappa/2$ respectively (cf Figure 2.4). There can exist zero, one or two ILRs, according to the pattern speed. The nature of stellar orbits in a bar potential have been computed by Contopoulos & Papayannopoulos (1980). They are gathered in families around periodic orbits, and they are either parallel or perpendicular to the bar, they change by 90° at each resonance. From the center to the corotation, if there is no ILR, orbits are parallel to the bar and support it, they are the x1 family. If there exist two ILRs, orbits between the two ILRs are perpendicular to the bar, and are called the x2 family. At corotation, the orbits in the rotating frame do not rotate around the center, but around the Lagrange points L4 and L5, on the minor axis of the bar. After the corotation, the orientation of the orbits changes again, and they become perpendicular to the bar and do not support it. The bar must therefore end at its corotation (see Figure 2.5 left).

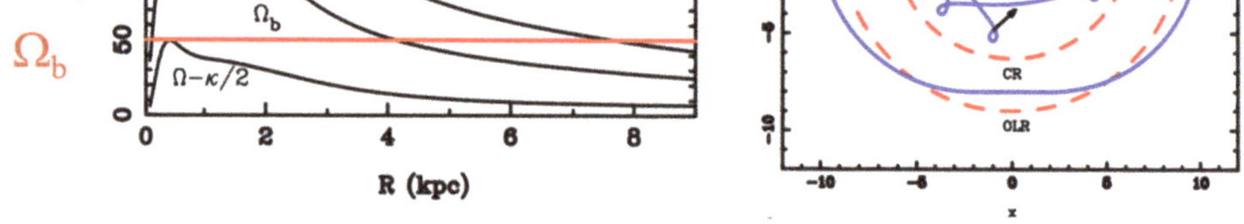

Figure 2.4. Left: rotation curve $V_{\rm rot}$ and associated frequencies Ω, $\Omega \pm \kappa/2$, for a typical spiral galaxy disk. A possible bar pattern speed Ω_b is traced in red, to delimitate the location of Lindblad resonances. Right: typical stellar orbits in this potential, where the three resonances (ILR, CR, and OLR) are traced by red-dashed circles, and some orbits in the rotating frame of the bar are traced in blue.

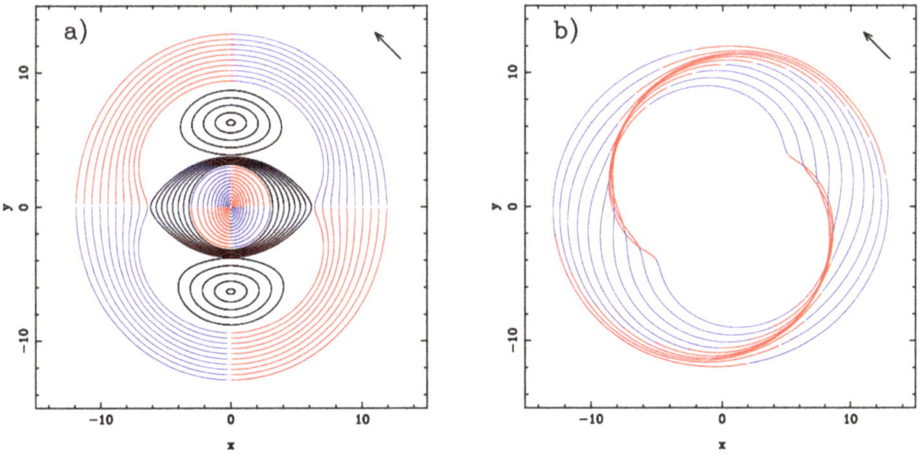

Figure 2.5. Left: schematic representation of the different stellar orbit families in a barred galaxy, where ILRs exist. Orbits are parallel to the horizontal bar from ILR to corotation, then perpendicular. Right: the gas in the bar potential flows with similar orientation, but the orbits precess due to cloud collision, and they turn gradually by 90° at each resonance. This produces orbit crowding, and the formation of a spiral wave. The orbits have been differently colored by quadrant, to better see the distribution in spiral arms.

2.2.3 Gravity Torques in a Barred Potential

The barred potential of the stellar component can induce spiral density waves in the gas component. The gas cannot follow the same periodic orbits as the stars; these periodic orbits can cross each other. Stellar orbits can cross since stars are a collisionless medium, but the physics of the gas is different. Gas clouds collide a few times per rotation, so their streamlines cannot cross but they are deflected. Instead of an abrupt jump by 90° at each resonance, the gas streamlines turn gradually to finally become perpendicular at each resonance. The gas response therefore has a spiral structure, whereas this spiral disappears with old stars (see Figure 2.5 right).

The fact that there is a phase-shift between the stellar potential and gas response, will create tangential forces and torques. The sign of the torques can be understood easily in the schematic representation of Figure 2.6. The bar separates in four quadrants the plane of the galaxy disk: two opposite quadrants (NE and SW) where the torque is positive, and two other quadrants (NW and SE) where the torque is negative relative to the direction of rotation of the galaxy, indicated by the arrow. The torques are symmetrical with respect to both minor and major axes of the bar. So if the gas was distributed symmetrically with respect to these axes, the torques will cancel out, averaged on one orbit/streamline. But the gas is distributed in a spiral structure, crowding in the two opposite quadrants of the same sign. Tangential forces from the bar will produce a non-canceling torque, transferring angular momentum.

The torques change sign at each resonance, as is easy to see geometrically in Figure 2.6. Outside of the corotation, the gas receives angular momentum from the bar and is driven outward to the outer Lindblad resonance (OLR), where it forms a

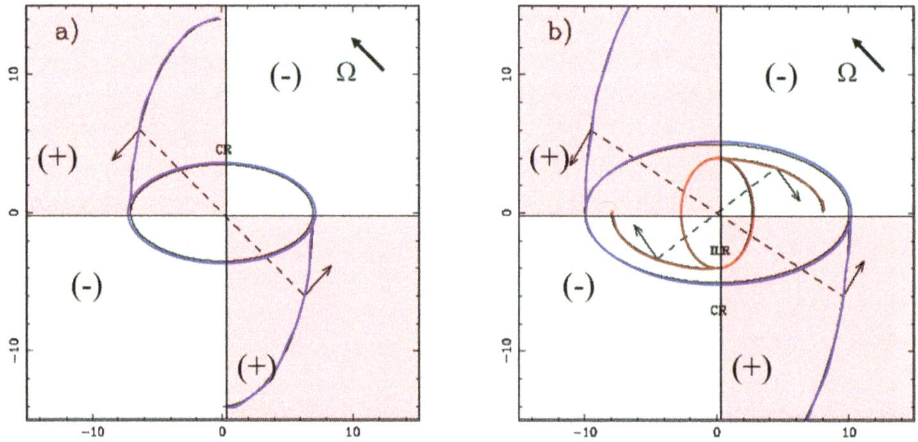

Figure 2.6. Schematic diagram showing geometrically the direction of the gravity torques exerted by the bar on the gas following a spiral structure: (a) outside of the corotation which corresponds roughly to the end of the bar, the torque is positive (red quadrant), and the gas is driven out toward the outer Lindblad resonance (OLR); (b) inside of the corotation, the torque is negative (green quadrant), and the gas will spiral toward the center, down to the ILR.

somewhat elongated ring (perpendicular to the bar just before the OLR, in the shape of a figure of eight; or elongated parallel to the bar just outside of the OLR). When it arrives at the resonance location, the spiral structure winds up in a ring, or pseudo-ring at the beginning. Since this morphology is symmetric with respect to the minor and major axes of the bar, the gravity torque decreases ad cancels out, the gas is stable in the ring. Inside corotation, the gas loses angular momentum and spirals toward the center. If there is an ILR and only one, the gas stops there and forms a nuclear ring symmetric with respect to the bar, where the torque averages out. Inside the ILR, the torque changes sign again, becomes positive between the center and ILR, carrying the gas outward to the ILR, where it accumulates and forms stars.

The winding sense of the spiral, depends on the shape of the precession rate $\Omega - \kappa/2$ of the elongated orbits. When the gas is infalling from corotation to the ILR, it sees the precessing rate increasing, and the gas is in advance at smaller radii, implying a trailing spiral. Once on the other side of the ILR, it sees in contrast the precessing rate decreasing, and it forms a leading spiral structure. This explains the reversal of the gravity torque. When the gas infalls within the sphere of influence of the central black hole, then the $\Omega - \kappa/2$ increases rapidly towards the center. This is the same configuration as the region between ILR and corotation, at larger radius, and the spiral becomes trailing again. In these case, the sign of the torque reverts again, and the gas is driven inward to feed the nucleus (cf Buta & Combes 1996). These two different cases are represented in Figure 2.7.

When the gas has been stalled at ILR, triggering star formation in hot spots distributed in a ring, without entering the sphere of influence of the black hole, the next step requires the formation of a nuclear bar, to take over the gas infall to the

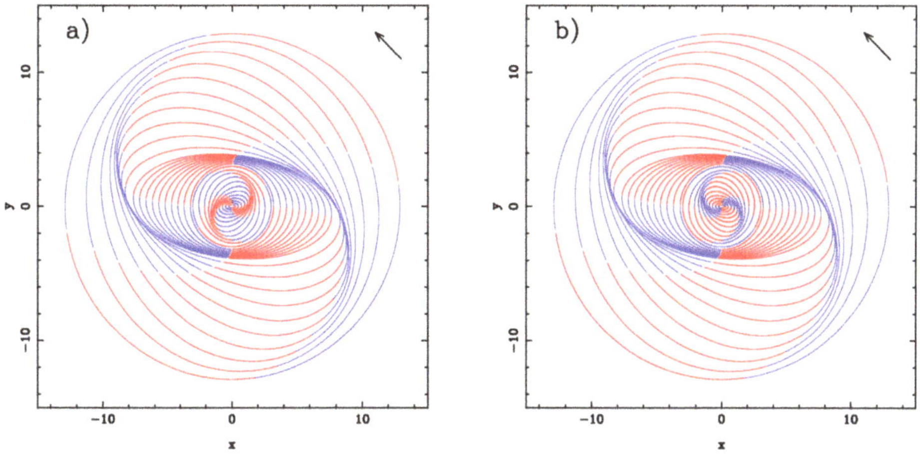

Figure 2.7. Left: schematic representation of the gas streamlines in a barred galaxy, where ILRs exist. The bar is horizontal. When there is no central mass concentration, and the precessing rate $\Omega - \kappa/2$ decreases towards the center, the gas is crowding in a leading nuclear spiral, and the torque exerted is positive. Right: when there is a supermassive black hole dominating inside the ILR, the precessing rate $\Omega - \kappa/2$ increases again toward the center, and the dissipating gas follows now a trailing spiral. The torque is negative, and the gas may fuel the nucleus.

nucleus. With the infall of new gas at ILR, forming new stars, the mass concentration gradually increases, which also increases the precession frequency $\Omega - \kappa/2$ and generates two ILRs. Since the periodic orbits inside the two ILRs are perpendicular to the bar, they weakened the bar and favor the apparition of a secondary or nuclear bar, decoupling inside the ILR region. This bar, supported by matter with high precession rate at the center, rotates with a higher pattern speed than the primary bar. In numerical simulations, it is often possible to observe a nuclear bar that shares resonances with the primary bar in order to minimize chaotic orbits. For example, the ILR of the primary bar is the corotation of the secondary bar (Friedli & Martinet 1993; Garcia-Burillo et al. 1998).

The nuclear bars extend the action and gravity torques of the primary bar to the center of the galaxy. They are the next step to feed active galaxy nuclei (e.g., Shlosman et al. 1989). In zoom-in numerical simulations, where a large range of scales can be taken into account, the gas infall is favored by a whole cascade of nested waves, either bars (Fourier mode $m = 2$) or lopsided waves ($m = 1$; Hopkins & Quataert 2010).

2.2.4 Observations of Nuclear Spirals

In recent high spatial resolution results, both with Atacama Large Millimeter Array (ALMA), and with SINFONI at the Very Large Telescope (VLT) at ESO, examples of these trailing nuclear spirals in nearby galaxies, within the sphere of influence of the black hole (20 pc-scales), have been uncovered. With ALMA, the molecular gas has been mapped at millimeter and sub-millimeter wavelengths, traced by emission lines of the CO molecule, and dense tracers like HCN and HCO+ (Combes et al.

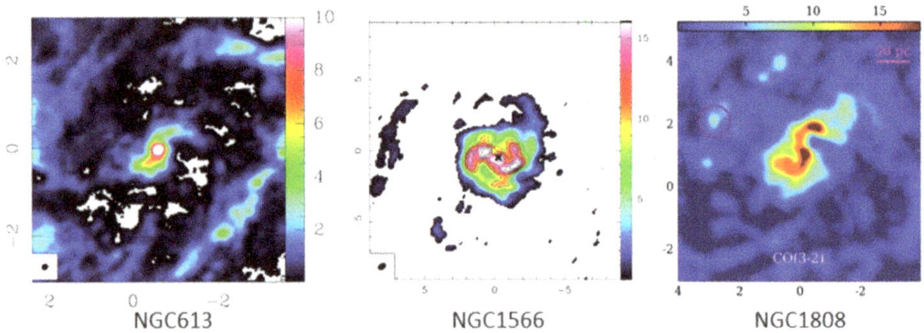

Figure 2.8. Three spiral galaxies, with a low-luminosity AGN, recently mapped with ALMA in the CO(3–2) emission line: NGC 613 (Audibert et al. 2019), NGC 1566 (Combes et al. 2014), NGC 1808 (Audibert et al. 2020). All reveal a nuclear spiral in molecular gas, inside a nuclear ring at the inner Lindblad resonance. Reproduced with permission from Audibert et al. (2019, 2020) and Combes et al. (2014). © ESO.

2014, 2019; Audibert et al. 2019, 2020). With SINFONI several near-infrared emission lines such as Brγ, [Fe II], Paα, H_2 yield complementary diagnostics (Falcon-Barroso et al. 2014; Smajic et al. 2015; Busch et al. 2017; da Silva et al. 2020).

The first case in Figure 2.8, NGC 613 is a strongly barred Seyfert galaxy with a ring at ILR, the locus of hot spots of star formation, like beads on a string with clumps of Brγ and CO/HCN emission (Falcon-Barroso et al. 2014). Inside this relatively empty ILR, there is a nuclear trailing spiral at the center, with radius ~50 pc. The spiral develops in a region where the black hole gravity dominates, and it is trailing (Audibert et al. 2019). Inside the nuclear spiral, the high resolution of ALMA reveals a kinematically decoupled molecular disk, or torus, of radius 14 pc (Combes et al. 2019). This torus appears to be fueled by the torque on the spiraling gas, according to the potential modeling.

NGC 1566 is a Seyfert-1 barred galaxy, showing a nuclear ring at the ILR of the primary bar, with a radius of ~1.25 kpc. Inside this ILR, there is a nuclear bar of about the same radius (1.2 kpc) with an ILR ring at R ~200 pc. The ALMA field of view is marginally able to see this ring at the border, but inside this secondary ILR, it uncovers a nuclear disk of about 85pc radius, which develops a trailing nuclear spiral. The latter develops from zero to 40pc radius, a region where the black hole gravity becomes dominant in the inner parts, explaining the trailing character of the spiral. This is a good example of embedded bars within bars, and resonances in cascade (Combes et al. 2014, 2019). A broad Brγ line witness and active nucleus and a 1000K black body emission a hot torus in the center (Smajic et al. 2015).

Finally, NGC 1808 is a starburst Seyfert-2 barred galaxy, with a galactic wind, seen as vertical dust lanes parallel to the minor axis, with the HST. The strong bar has an ILR corresponding to a ring of star formation at a radius of 450 pc. A very contrasted trailing nuclear spiral up to 50 pc radius is seen in CO emission, but also in HCO+, HCN, and CS. Inside the spiral, ALMA reveals a kinematically decoupled molecular torus of 6 pc radius (Audibert et al. 2020). There is no molecular outflow near the nucleus, although there is one at 100pc scale (Busch et al. 2017), which must come from the supernovae feedback. The stellar potential given by a nuclear bar, of radius 180pc, exerts negative torques on the gas in the trailing nuclear spiral, which is feeding the embedded molecular torus.

In all these objects, the precise knowledge of the potential from the near-infrared image at high resolution, after proper deprojection and calibration of the mass-to-light ratio using the rotation curve, allows us to measure the gravity torques in the galaxy plane, and determine the four quadrants. The molecular gas map then indicates whether these torques are actually exerted on the gas, and what is in average the exchange of angular momentum with the bar/nuclear bar. In all cases, the computations resulted in gas inflow, feeding the nucleus. Note that this inflow is only indirectly derived, it is not possible to directly measure the inflow velocities, of the gas in rotation equilibrium in the disk, even with non-circular motion.

2.3 AGN Feeding during Galaxy Interactions and Mergers

Galaxy interaction and mergers can enhance the inflow of material from galactic scales into the close environments of active nuclei. Gravity torques are important during interactions, since the total potential is strongly non axisymmetric. The tidal forces due to a companion are bisymmetric ($m = 2$) when the relative distance is large with respect to galaxy radii, and then it becomes more complex, with lopsided components ($m = 1$) during the merger. Already at a distance, the tidal forces can trigger the formation of a bar, through their bisymmetric shape: there is first material arms (bridge and tail), which develop through the SWING mechanism to form a self-gravitating feature (Toomre 1981), and a strong bar, as conformed through simulations (Barnes & Hernquist 1991, 1996). The strong bar then produces gravity torques that may drive the gas towards the center, to fuel both a starburst and an active nucleus.

Major mergers could be the essential way to fuel the most luminous quasars (e.g., the review by Storchi-Bergmann et al. 2019). Already it has been found through observations that the most extreme starbursts, the ultra-luminous infrared galaxies, or ULIRGs, are always mergers of gas-rich galaxies (e.g., Sanders & Mirabel 1996). During the starburst, and at the end phase, it is frequent to find a luminous quasar, and quasars are observed with a young stellar population (Kauffmann et al. 2003). This has built the scenario where the major merger of two gas-rich spiral galaxies drive large amounts of gas to the center, producing nuclear starbursts, and a lot of obscuration. In parallel, the gas is feeding a supermassive black hole, but so obscured that it will manifest itself clearly only at the end, when feedback will expel the gas and uncovers a bright optical quasar (Hopkins et al. 2006). After the feedback phase, the quasar luminosity decreases by lack of fuel, and the object becomes a quiescent galaxy. It is possible to simulate the right amount of feedback, to derive the duty cycle of the AGN at a certain luminosity, and derive the luminosity function of quasars to be compared with observations (cf Figure 2.9). Only major mergers can explain the most luminous quasars. However, most of the other AGN can be accounted for by secular evolution or minor mergers (Hopkins et al. 2008).

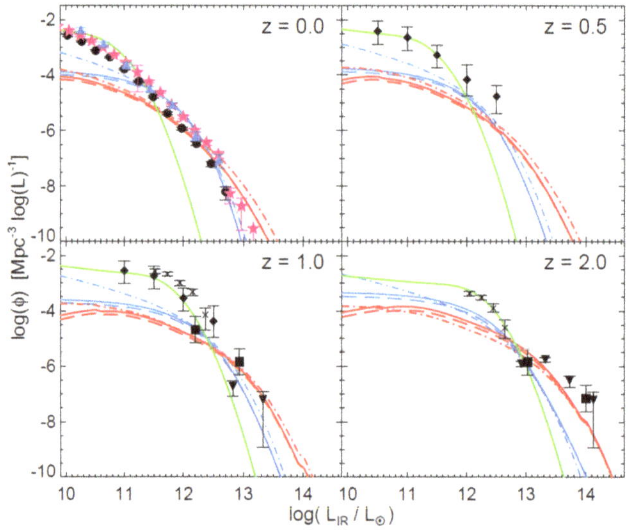

Figure 2.9. Comparison between observed (points and error bars) and simulated (colored lines) total IR (8–1000μm) luminosity functions at different redshifts ($z = 0, 0.5, 1$ and 2). Green, blue, and red lines show the estimated contribution from non-merging systems, star formation in mergers, and obscured AGN in mergers, respectively. Dashed and dotted lines correspond to different parameters of the model. Reprinted from Hopkins et al. (2008).

The scenario is able to explain the downsizing observed phenomenon, i.e., that most very luminous quasars and starbursts occur in the past, in the first half of the universe, and only lower-luminosity and less-massive AGN and star formation are now on-going. It can also explain that the star formation and black hole growth are statistically concomitant across cosmic time. The rates of star formation and central black hole accretion follow a similar rise and fall, with a peak at $z = 1.9$, 3.5 Gyr after the Big Bang (Madau & Dickinson 2014). If major mergers are the main sources of the most luminous quasars, the latter are not numerous, and the total luminosity fraction contributed by these mergers, both in star formation and black hole accretion is minor (5%–20%, according to redshift). The bulk of the luminosity comes from more normal mass galaxies and secular evolution (Michalowski et al. 2012, Bourne et al. 2017).

From the AGN luminosity function and its redshift evolution, it is possible to derive the supermassive black hole mass function and its growth rate, since the source of energy of nuclei is the accretion onto black hole, given assumptions about the radiative efficiency and the Eddington ratios. This argument was first used by Soltan (1982) to estimate at 8×10^{13} $M_{\odot}$ the mass of black holes per Gpc^3 today, and has widely been used since. The total density of black holes is now estimated ~6 times higher, at 3–5.5 $\times 10^5$ $M_{\odot}$/Mpc^3, and more details about AGN duty cycles, and the epoch of black hole growth can be derived (e.g., Shankar et al. 2009). Nuclei were keeping active more often, about 7% of their life in the early universe, while today the activity has slowed down to 0.4%. Following the downsizing, the Eddington ratios decrease with time, and also with mass, meaning that high mass

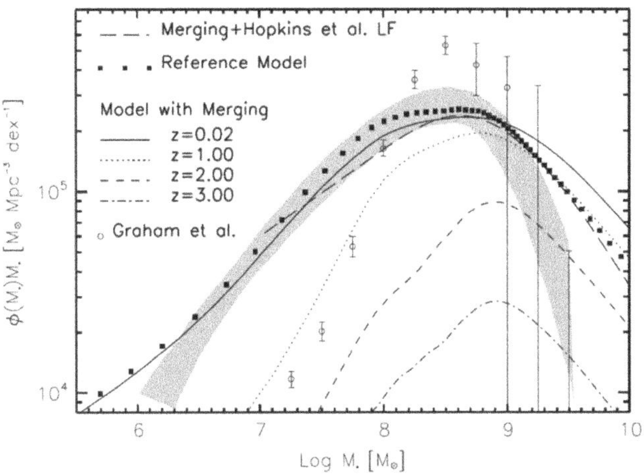

Figure 2.10. The $z = 0$ observed black hole mass function (gray band) is compared with different evolution models, including black hole mergers. The long-dashed line shows the $z = 0$ mass function for the merger model of Hopkins et al. (2007) and open circles show the determination of Graham et al. (2007) using the relation between black hole mass and Sersic index. The growth models through accretion at different redshifts are computed assuming an average radiative efficiency and initial Eddington ratio, and each black hole has a probability of 50% of merging with another black hole of equal mass per Hubble time. Squares show the $z = 0$ predictions of the reference model (same accretion parameters, no merging). Reprinted from Shankar et al. (2009). © 2009. The American Astronomical Society. All rights reserved.

black holes were formed rapidly in the first half of the universe, may be through mergers. More low-mass black holes are still forming now, in low-luminosity AGN. The distribution of black hole mass is well described by the mass function $\Phi(M_{\rm BH})$, multiplied by the mass $M_{\rm BH}$, as displayed in Figure 2.10.

The results of the models shown in this figure reveal that, although some parameters are kept fixed, the range of viable models is still huge, and more observations are required to restrain the freedom. The observation of higher black hole masses with respect to bulge mass at high redshift (Wang et al. 2013; Venemans et al. 2016), may require a variation of the fixed parameters, in order to allow some decoupling between stellar mass and black hole mass growth in galaxies. Recent studies based on the X-ray luminosity function find a BH-growth non-linear with the Galaxy-host growth (Delvecchio et al. 2020; Ananna et al. 2020), taking into account about 50% obscured (Compton thick) AGN.

2.4 Formation at High Redshift: What Kind of Seeds?

2.4.1 The Growth Rate

The problem to grow black holes rapidly enough has always been a difficult issue, which is even more severe today, that several z~6–7 quasars, with black hole mass larger than $10^9\ M_\odot$ have been discovered (e.g., Fan et al. 2001). Indeed, the rate of growth of a given black hole is proportional to its mass. This is due to the limiting factor of the radiation pressure. If the accretion rate is too large, the gas radiates a maximum, and the radiation pressure force will be larger than the gravity force, which is proportional to the mass. The maximum accretion rate, is called the Eddington rate, and is proportional to the mass. $L_{\rm Edd} = 3.3\ 10^{12} L_{\odot}\ (M/10^8 M_{\odot}) = 0.1\ c^2\ dM/dt$, with the usual efficiency of 10%. This implies, for a constant accretion rate equal to the maximum, an exponential growth $dM/dt = M/\tau$ or $M = M_0 \exp(t/\tau)$, with a characteristic e-folding timescale of $\tau = 45$ Myr. This timescale looks relatively small, however the seed to begin the black hole growth is only at the stellar-mass scale, typically 10 $M_{\odot}$. Then it will require 860 Myr at the Eddington rate to reach $2 \times 10^9\ M_{\odot}$. However, from redshift $z = 20$, where the first stars and dwarf galaxies begin to form, and $z = 7$, there is only 600 Myr, and not enough time to grow the observed BH. At $t = 600$ My, the maximum BH masses are $6 \times 10^6\ M_{\odot}$.

One solution that has been proposed recently is that the first black holes accrete beyond the maximum, at several times the Eddington rate. This might be possible, since the Eddington estimation has been done with the assumption of spherical symmetry. It is possible that some gas is inflowing in 2D in a plane, and is shielded from the AGN radiation through self-absorption or obscuration by dust. The wind pushed by radiation pressure will get out in the perpendicular direction. MHD simulations have shown that it is still possible to accrete, although at a reduced rate, given the outflow, beyond a few hundreds of Eddington rate (Jiang et al. 2019). Also, the ultra-luminous X-ray sources (ULX), whose luminosities are beyond the Eddington rate of stellar-mass black holes, appear to be in many cases pulsars radiating at super-Eddington rates, so instead neutron stars and not black holes (Bachetti et al. 2014; Israel et al. 2017). In this context, it is possible to divide the typical e-folding time by a factor 10 or 100, and there is no more problem to form black holes rapidly from stellar seeds. Maintaining the accretion rate at such super-Eddington rates at high redshift is not easy however.

2.4.2 Different Modes of Accretion Versus Redshift

Very massive black holes, with masses 2–6 $\times 10^9$ solar masses, are present less than 1 Gyr after the Big Bang. More than a dozen quasars at $z = 6$–7 have been observed both in broad optical lines, and in their molecular content via CO lines (e.g., Venemans et al. 2016). The width of the optical lines, combined with an estimation of the size of the BLR (Broad Line Region) yields the mass of the black hole, through the virial relation. The size of the BLR is derived from the empirical relation between the bolometric luminosity of the quasar and the BLR size, obtained through the reverberation mapping studies (Peterson & Bentz, 2006). The mass of the host galaxy (or the bulge mass) is derived from the kinematics of the CO lines, assuming an inclination of the galaxy disk. The results show that black holes in the early universe are more massive than expected for the scaling relation obtained locally, and by sometimes one or two orders of magnitude (cf Figure 2.11). It is possible in early times that black hole growth is faster than galaxy growth, and star formation will catch over later. Cosmological simulations have found such fast growing black holes in very massive halos of $5 \times 10^{12}\ M_{\odot}$, at the crossing of filaments. They are

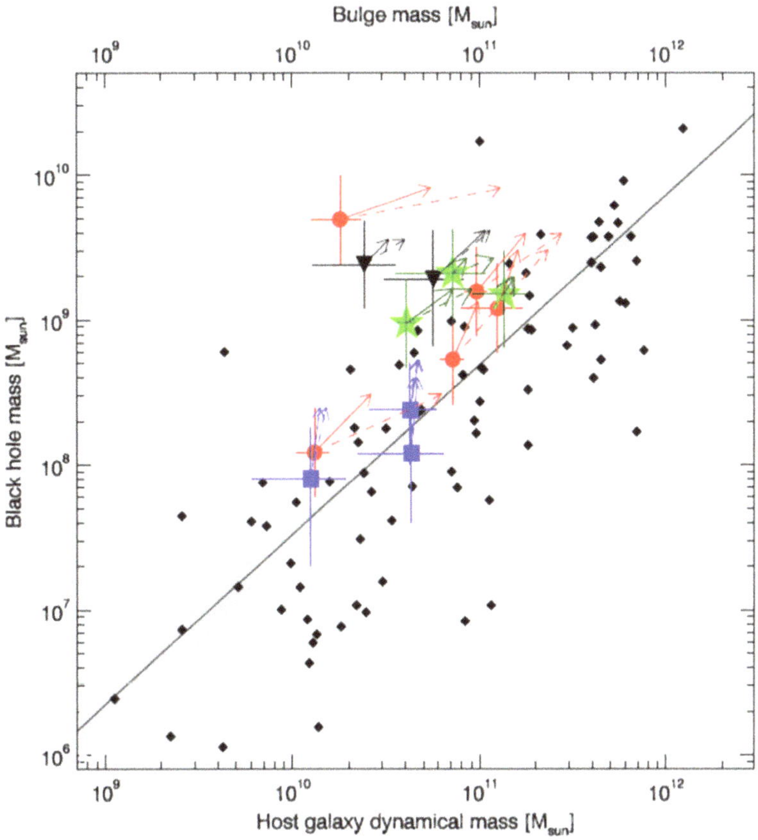

Figure 2.11. The black hole mass versus the dynamical mass of z~6 quasar host galaxies and the bulge mass of local galaxies. The black diamonds are values obtained for local galaxies (Kormendy & Ho, 2013). Their MBH–M_{bulge} relation is represented by the solid line and the shaded area. The large and colored symbols are the high-z quasars. The green stars are the z>6.5 quasars from Venemans et al. (2016). For a given bulge mass, the high-redshift quasars have a more massive black hole than local galaxies. From the quasar luminosity (linked to its accretion rate), and from the observed star formation rate, it is possible to extrapolate the trajectory of the points (arrows) during the next 50 Myr. Reprinted from Venemans et al. (2016).

fueled exceptionally fast, and these quasars will become cD galaxies, the brightest galaxies at the center of rich clusters of $2 \times 10^{15}\ M_{\odot}$ (e.g., Bhowmick et al. 2020).

Semi-analytical models have been used to investigate how the dominant mechanisms of black hole growth are changing with redshifts, given the observational constraints. As a general feature, the growth is dominated by gas accretion until z~1, and then by black hole coalescence during galaxy mergers in the second half of the universe, when abundant gas is less available. More precisely, as a function of luminosity, the dominant accretion mechanisms at redshift lower than 2 is through ADAFs for low-luminosity AGN (see Figure 2.12), while it is through thin disks, at z>2, and super-Eddington accretion, especially for the high-luminosity end (Griffin et al. 2019).

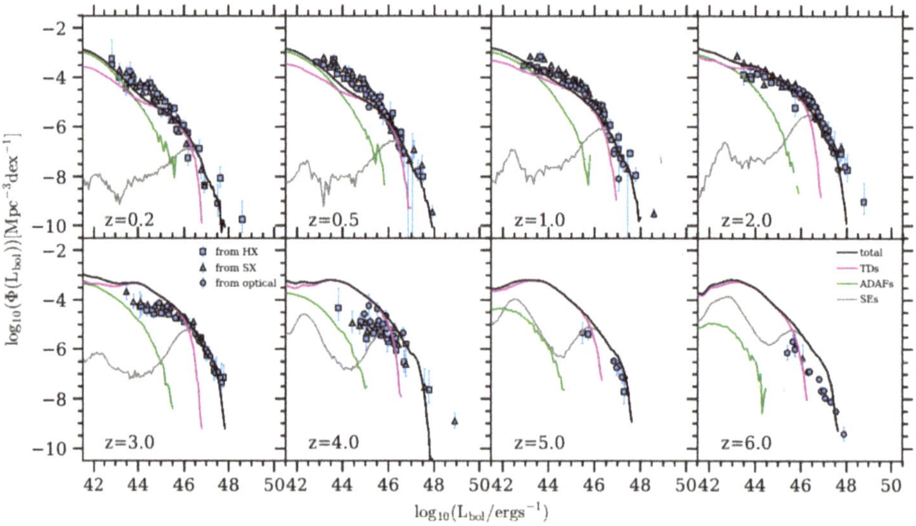

Figure 2.12. Computation through a semi-analytical model of the various contributions to the AGN luminosity function, as a function of redshift. The total bolometric luminosity function of the model (black line) is compared to the observation points, indicated by the wavelength (squares—hard X-ray, triangles—soft X-ray, circles—optical). The black line is the sum of the contribution of the ADAF mode (green), thin disks (purple), and super-Eddington objects (gray). Reproduced from Griffin et al. (2019). © 2019. The Royal Astronomical Society. All rights reserved.

The formation of high-z very massive black holes poses the problem of the seeds, related to the intermediate-mass range of black holes (IMBH), that are very difficult to detect, even locally. Given their rarity, some hypotheses have been proposed that black holes can form through direct collapse of gas clouds, bypassing the stage of IMBH (Shang et al. 2010). However, some IMBH have been detected in dwarf galaxies, in particular through their AGN manifestations. Searches in globular clusters have not been successful.

2.4.3 Intermediate Mass Black Holes

IMBH, between 100 and 10^5 $M_\odot$, should normally be the inevitable stage between stellar-mass black holes and the SMBH seen in almost all giant galaxy nuclei. However, they are difficult to detect, first due to their small mass, but also because they are not necessarily located at the center of their galaxy host. Dynamical friction has a timescale inversely proportional to the black hole mass, and IMBHs might not have had enough time to spiral in since their formation. One of the first IMBH to be detected in a dwarf galaxy was in NGC 4395 (Filippenko & Ho 2003), a bulgeless Seyfert-1 galaxy, with $M_{\rm BH} = 3 \times 10^5$ $M_\odot$, a low ionization, and an Eddington ratio of 2×10^{-3}–2×10^{-2}. An important problem to detect kinematically IMBH in nuclei is that dwarf galaxies frequently have their gravity dominated by a nuclear cluster, of mass of the order of 10^6 $M_\odot$. Although these dwarf galaxies are bulgeless, a scaling relation has been obtained with the total stellar mass M_*, at least within a morphological class, the slope is comparable to the usual scaling relation. More

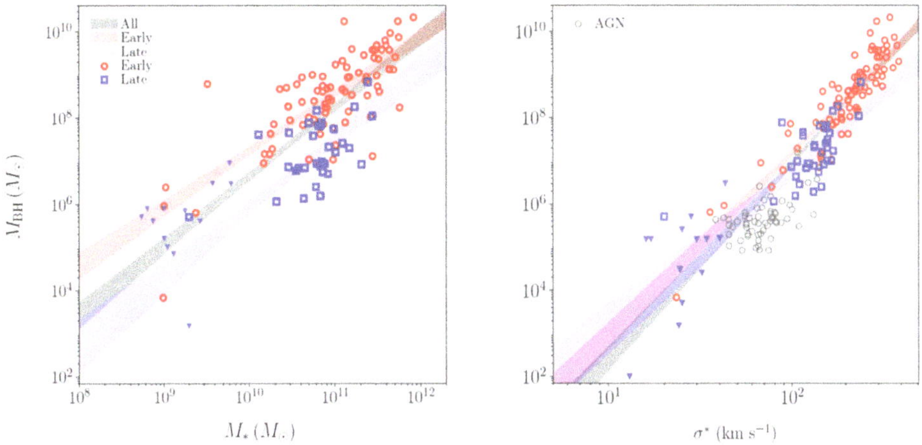

Figure 2.13. Although small-mass galaxies are bulgeless, they still follow the M–σ relation established for the giant galaxies. The left panel shows the $M_{\rm BH}$ versus M_* for dynamical early-type (red open circles), late-type (blue open squares), and dynamical upper limits (blue triangles). The colored bands correspond to the fits for early and late-type galaxies (red and blue shaded regions) and the full sample (gray). Although all fits have the same slope, they differ by their zero points. The right panel shows the $M_{\rm BH}$ versus σ_*, where the fits are more comparable. The gray dots are the active galactic nuclei sample from Xiao et al. (2011). Reprinted from Greene et al. (2020). © 2020 Annual Reviews.

interestingly, they follow even better the M–σ relation, although with much scatter, as shown in Figure 2.13, from the recent review by Greene et al. (2020). The very nearby example of M33 is still an exception, where no IMBH has yet been found, with an upper limit of $10^3\ M_\odot$, a factor 10 below the scaling relation. Although M33 possess a nuclear stellar cluster, which in principle favor the formation of black holes, it is possible that the merger of two IMBH in two nuclear clusters has produced a recoil of the seed black hole out of the Galaxy (e.g., Askar et al. 2021).

More than three hundred IMBH candidates were identified in a sample of a million spectra from SDSS galaxies by Chilingarian et al. (2018). The candidates were selected through their emission line ratio diagnostics (BPT diagram) pointing towards AGN excitation. In ten sources, the AGN was confirmed through X-ray emission. These detections, although rare, show that nuclear IMBH are the low-mass extension of the scaling relation between black hole and bulge masses in galaxies.

Ultra-luminous X-ray sources were first thought to be candidates for IMBH, and now have revealed to be super-Eddington accreting stellar-mass black holes, or even neutron stars, as mentioned in section 4.1. However, there exist hyper-luminous X-ray sources that are still good candidates for IMBH. The prototype is HLX1, in the ESO 243–49 galaxy system, at a distance of 95 Mpc. Its luminosity is 10^{42} ergs s^{-1}, corresponding to the Eddington luminosity for an $M = 10^4\ M_\odot$. The mass of the black hole is likely to be between 10^2–$10^5\ M_\odot$ (Farell et al. 2009, Webb et al. 2010). HLX1 happens to be hosted by a dwarf galaxy, in interaction with the lenticular galaxy ESO 243–49. Hα spectroscopy has shown that the two companions are

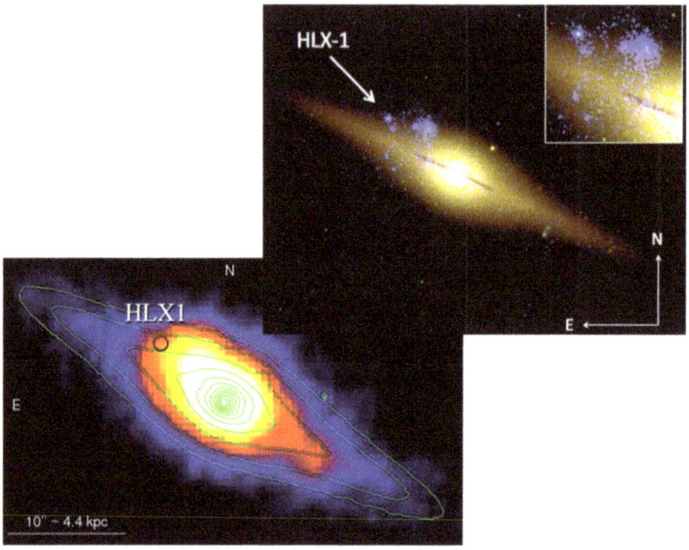

Figure 2.14. Images of one of the best IMBH candidate, HLX1, hosted by a dwarf galaxy, companion of the lenticular galaxy ESO 243–49. At top is an optical HST photograph (Mapelli et al. 2013), and at bottom, an UV at 2000 Angstrom image (Swift/UVOT, Webb et al. 2010, Soria et al. 2013), with R-band contours superposed. Reprinted from Webb et al. (2010).

dynamically associated (Soria et al. 2013). There has been X-ray flares, corresponding to quasi-periodic outbursts every year, with a last one delayed by one month, may be due to a star in orbit around the IMBH losing some mass (Godet et al. 2014). There is star formation detected in projection towards HLX1, coming from the tidally disrupted dwarf galaxy (cf Figure 2.14).

The lenticular galaxy and its dwarf companion belong to the Abell 2877 cluster, and they are subject to the ram pressure from the intracluster medium. Mapelli et al. (2013) run a set of N-body simulations showing that a minor merger between the two galaxies (mass ratio 20:1) is able to explain the observations, provided that the first pericenter occurred more than 2.5 Gyr ago. The merger triggered star formation in the dwarf nucleus, at the origin of a star cluster much more massive than 10^4 solar masses, in which the IMBH could have formed.

2.4.4 Black Holes in Globular Clusters

One of the most appealing way of IMBH formation is in globular clusters. These very dense clusters have a relaxation timescale shorter than their age, due to their relatively small total number N of stars (see Section 2.1). Due to two-body relaxation, energy is transferred from the center to the outer parts, triggering a gravothermal catastrophe and the core collapse in about 10 relaxation time, i.e., in less than 1 Gyr. While the core is denser and denser, stellar binaries interact with other stars and other binaries, delivering energy, when the binaries tighten, which

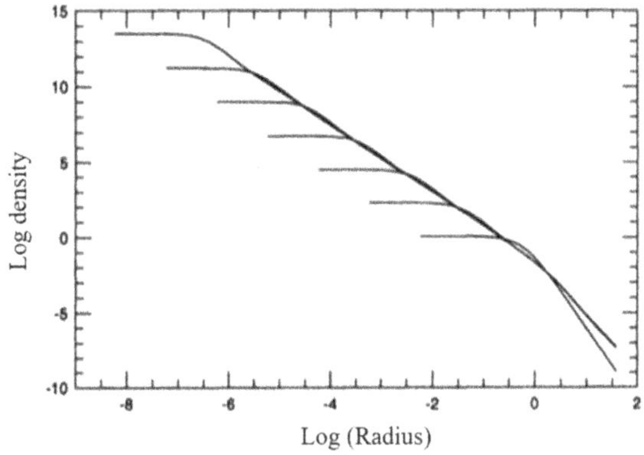

Figure 2.15. Schematic drawing of the density evolution of a globular cluster, leading to the core collapse. Two-body relaxation transfers energy from the center to the outer parts, and the density increases up to a run-away, a gravothermal catastrophe towards a black hole collapse.

can delay the core collapse for some time. At the beginning of the evolution, the cluster is in equilibrium with a radial King profile, with a flat core in the center (Figure 2.15). The relaxation tends towards energy equipartition, the massive bodies infall towards the center, while the light ones evaporate (mass segregation). While becoming denser, the core size reduces considerably, and a core-collapsed cluster is then characterized by a power law slope in the center, instead of a core (Chatterjee et al. 2013).

Although some globular clusters are observed in the category of core-collapsed, it has been difficult to detect any IMBH in their center. Recently Vitral & Mamon (2021) have failed to detect the signature of IMBH in the center of one of the best candidates, the nearby core-collapsed, globular cluster NGC 6397. They used 3D velocities (proper motions from HST and GAIA), a sophisticated dynamical model, and confirmed the existence of a central dark component, but more in the shape of a diffuse sub-cluster of unresolved objects than a point source. If an IMBH exists in the center of a globular cluster, it should scatter the orbits of stars, and quench the mass segregation, established by the two-body relaxation. Resolving stars in the nearby cluster NGC 2298 by HST photometry, Pasquato et al. (2009) have put a limit on the mass of a possible black hole at 1% of the cluster mass. In conclusion, there is no evidence of IMBH in globular clusters. However, the recent gravitational wave event GW190521 has demonstrated that the merger of black holes can produce objects entering the IMBH mass range. This event corresponds to the merger of a 65 $M_\odot$ with an 85 $M_\odot$ black hole yielding a 142 $M_\odot$ at the end. All these black holes belong to the mass range between 60 and 130 $M_\odot$ where stellar-mass black holes are not expected. Indeed, a very massive star exploding at the end of its life, would make a supernovae unstable by pair creation; the residual after explosion will pulse, and losing pressure support will collapse violently and disperse all the material in the interstellar medium. It is very likely that collisions between stars in dense clusters lead to multiple successive BH mergers, and this is one of the privileged way to form IMBH (Kremer et al. 2020).

2.4.5 Direct Collapse—Quasi-stars

To form very massive black holes at high redshift, several ways have been proposed, to start with an already massive seed. The first stars are certainly very poor in metals, being composed of primordial gas, with only light elements from the Big Bang. Lacking strong coolants, the giant gas clouds cannot fragment, and the Jeans mass is much more massive than today. This first generation of stars, called PopIII stars, are much above the limit mass of local stars, and could be massive seeds for black holes. These stars have a very short lifetime, and their explosion contribute to enrich the interstellar medium in metals, so that the next generation of stars will be less massive. Typically, PopIII stars of ~10^3 $M_\odot$, explode in supernovae, with a residual of ~10^2 $M_\odot$. Even more efficient will be the direct collapse of cold primordial clouds in a black hole, or the formation of a quasi-star. Dense and massive gas clouds accumulate at the center of galaxies, which trigger the collapse of a supermassive star, with already the formation of a small black hole in its interior. The giant star keeps on accreting gas until the final total collapse to a massive black hole. The supermassive star could then have a mass larger than 10^6 $M_\odot$, and collapse into a black hole of mass larger than 10^4 $M_\odot$ (e.g., Schleicher et al. 2013). These huge objects, called quasi-stars, look like red giants (cf Figure 2.16). They have a convective envelope, supported by pressure. The photospheric temperature decreases with the growth of the black hole. With such massive seeds, it is possible to account for all massive black holes at z larger than 6, with accretion and mergers, provided that there is no ejection, during a three-body interaction. The slingshot effect, leading to the ejection of a massive black hole in the intergalactic medium, could be due to the interaction between three BH during a group merger or even during a two-BH merger, followed by anisotropic radiation of gravitational waves. Depending on both the relative orientation of the spins of the merging BHs and their mass ratio, the merged BH may receive a recoil kick (Centrella et al. 2010, Komossa 2012) with velocities as high as ~5000 km s^{-1}, ejecting the black hole from its galaxy host. This event is however rare, it has been observed in 3C186 (Chiaberge et al. 2017) and CID42 (Civano et al. 2010).

The direct collapse of gas clouds into a black hole requires quite specific conditions (e.g., Volonteri et al. 2010; Devecchi et al. 2012). To avoid cooling and fragmentation, the molecular hydrogen, main coolant when metals are absent, has to be photo-dissociated, through strong UV photons sent by a nearby galaxy. Also angular momentum should be transferred quickly to the outer part, through gravity torques in a disk.

In summary, there are at least three processes to form supermassive black holes very early in the universe: the seeds coming from the remnant of PopIII stars, stellar collisions in dense clusters, like globular clusters, and direct collapse of gas clouds in a massive BH. The probability of all these processes are comparable in black hole growth models, although the three can form different masses, up to $10^3 M_\odot$ for PopIII seeds, up to 10^4 $M_\odot$ for star clusters, and up to 10^6 $M_\odot$ for direct collapse (Devecchi & Volonteri 2009).

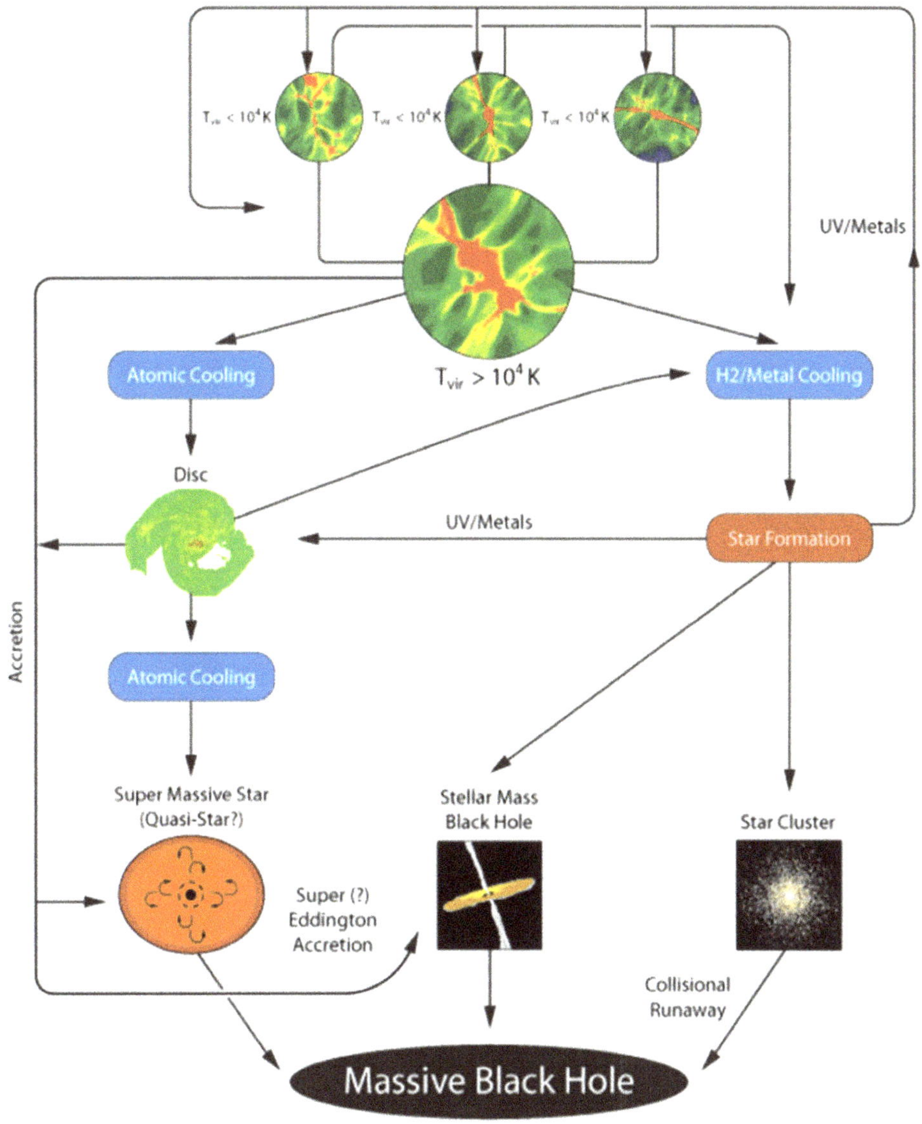

Figure 2.16. Summary of the possible pathways to massive black holes via a stellar seed black hole, a quasi-star or via a nuclear star cluster, in DM haloes with Tvir & 10000K. Reproduced from Regan & Haehnelt (2009).

Bibliography

Alexander, T. 2017, ARA&A, 55, 17
Ananna, T. T., Urry, C. M., Treister, E., et al. 2020, ApJ, 903, 85
Askar, A., Davies, M. B., Church,, & Ross, P. 2021, MNRAS, 502, 2682
Auchettl, K., Ramirez-Ruiz, E., & Guillochon, J. 2018, ApJ, 852, 37
Audibert, A., Combes, F., Garcia-Burillo, S., et al. 2019, A&A, 632, 33

Audibert, A., Combes, F., Garcia-Burillo, S., et al. 2021, A&A, arXiv: 20110.9133
Bachetti, M., Harrison, F. A., Walton, D. J., et al. 2014, Natur, 514, 202
Barnes, J. E., & Hernquist, L. E. 1991, ApJ, 370, L65
Barnes, J. E., & Hernquist, L. E. 1996, ApJ, 471, 115
Bhowmick, A. K., Somerville, R. S., Di Matteo, T., et al. 2020, MNRAS, 496, 754
Bourne, N., Dunlop, J. S., Merlin, E., et al. 2017, MNRAS, 467, 1360
Busch, G., Eckart, A., Valencia-S, M., et al. 2017, A&A, 598, A55
Buta, R., & Combes, F. 1996, FCPh, 17, 95
Cappellari, M. 2016, ARA&A, 54, 597
Centrella, J., Baker, J. G., Kelly, B. J., et al. 2010, RvMP, 82, 3069
Chatterjee, S., Umbreit, S., Fregeau, J. M., & Rasio, F. A. 2013, MNRAS, 429, 288
Chiaberge, M., Ely, J. C., Meyer, E. T., et al. 2017, A&A, 600, A57
Chilingarian, I. V., Katkov, I. Y., Zolotukhin, I. Y., et al. 2018, ApJ, 863, 1
Civano, F., Elvis, M., Lanzuisi, G., et al. 2010, ApJ, 717, 209
Combes, F., Garcia-Burillo, S., Casasola, V., et al. 2014, A&A, 565, A97
Combes, F., Garcia-Burillo, S., Audibert, A., et al. 2019, A&A, 623, A79
Contopoulos, G., & Papayannopoulos, T. 1980, A&A, 92, 33
da Silva, P., Menezes, R. B., Steiner, J. E., & Fraga, L. 2020, MNRAS, 496, 943
Delvecchio, I., Daddi, E., Aird, J., et al. 2020, ApJ, 892, 17
Devecchi, B., & Volonteri, M. 2009, ApJ, 694, 302
Devecchi, B., Volonteri, M., Rossi, E. M., et al. 2012, MNRAS, 421, 146
Dong, X.-B., Ho, L. C., Yuan, W., et al. 2012, ApJ, 755, 167
Emsellem, E. 2013, MNRAS, 433, 1862
Event Horizon Coll, 2019, ApJ, 875, L1
Faber, S. M., & Jackson, R. E. 1976, ApJ, 204, 668
Falcon-Barroso, J., van de Ven, G., Peletier, R. F., et al. 2011, MNRAS, 417, 1787
Falcon-Barroso, J., Almeida, R., Böker, C., et al. 2014, MNRAS, 438, 329
Fan, X., Narayanan, V. K., Lupton, R. H., et al. 2001, AJ, 122, 2833
Farrell, S. A., Webb, N. A., Barret, D., et al. 2009, Natur, 460, 73
Ferré-Mateu, A., Trujillo, I., Martín-Navarro, I., et al. 2017, MNRAS, 467, 1929
Filippenko, A. V., & Ho, L. C. 2003, ApJ, 588, L13
French, K. D., Wevers, T., Law-Smith, J., Graur, O., & Zabludoff, A. I. 2020, SSRv, 216, 32
Friedli, D., & Martinet, L. 1993, A&A, 277, 27
Garcia-Burillo, S., Sempere, M. J., Combes, F., & Neri, R. 1998, A&A, 333, 864
Genzel, R., Eisenhauer, F., & Gillessen, S. 2010, RvMP, 82, 3121
Godet, O., Lombardi, J. C., Antonini, F., et al. 2014, ApJ, 793, 105
Graham, A. W., Driver, S. P., Allen, P. D., & Liske, J. 2007, MNRAS, 378, 198
Graham, A. W., Onken, C. A., Athanassoula, E., & Combes, F. 2011, MNRAS, 412, 2211
Greene, J. E., Strader, J., & Ho, L. C. 2020, ARA&A, 58, 257
Greenhill, L. J., Jiang, D. R., Moran, J. M., et al. 1995, ApJ, 440, 619
Griffin, A. J., Lacey, C. G., Gonzalez-Perez, V., et al. 2019, MNRAS, 487, 198
Habibi, M., Gillessen, S., Pfuhl, O., et al. 2019, ApJ, 872, L15
Hills, J. G. 1975, Nat, 254, 295
Hopkins, P. F., Hernquist, L., Cox, T. J., et al. 2006, ApJS, 163, 1
Hopkins, P. F., Hernquist, L., Cox, T. J., et al. 2007, ApJ, 669, 45
Hopkins, P. F., Hernquist, L., Cox, T. J., & Keres, D. 2008, ApJS, 175, 356

Hopkins, P. F., Lauer, T. R., Cox, T. J., et al. 2009, ApJS, 181, 486
Hopkins, P. F., & Quataert, E. 2010, MNRAS, 407, 1529
Israel, G. L., Papitto, A., Esposito, P., et al. 2017, MNRAS, 466, L48
Jiang, Y.-F., Stone, J. M., & Davis, S. W. 2019, ApJ, 880, 67
Kauffmann, G., Heckman, T. M., Tremonti, C., et al. 2003, MNRAS, 346, 1055
Komossa, S. 2012, AdAst, 2012, 364973
Kormendy, J., & Ho, L. 2013, ARA&A, 51, 511
Kremer, K., Spera, M., Becker, D., et al. 2020, ApJ, 903, 45
Lin, C. C., & Shu, F. H. 1964, ApJ, 140, 646
Lin, D. N. C., & Pringle, J. E. 1987, MNRAS, 225, 607
MacLeod, M., & Lin, D. N. C. 2020, ApJ, 889, 94
Madau, P., & Dickinson, M. 2014, ARA&A, 52, 415
Magorrian, J., Tremaine, S., Richstone, D., et al. 1998, AJ, 115, 2285
Mapelli, M., Annibali, F., Zampieri, L., & Soria, R. 2013, MNRAS, 433, 849
McConnell, N. J., & Ma, C.-P. 2013, ApJ, 764, 184
Merritt, D. 2013, CQGra, 30, 244005
Michałowski, M. J., Dunlop, J. S., Cirasuolo, M., et al. 2012, A&A, 541, A85
Miller, J. M., Kaastra, J. S., Miller, M. C., et al. 2015, Nature, 526, 542
Neumayer, N., Seth, A., & Böker, T. 2020, A&ARv, 28, 4
Newman, A. B., Ellis, R. S., Andreon, S., et al. 2014, ApJ, 788, 51
Norman, C., & Scoville, N. Z. 1988, ApJ, 332, 124
Pasquato, M., Trenti, M., De Marchi, G., et al. 2009, ApJ, 699, 1511
Peterson, B. M., & Bentz, M. C. 2006, NewAR, 50, 796
Pfister, H., Volonteri, M., Dai, J. L., & Colpi, M. 2020, MNRAS, 497, 2276
Pfister, H., Dai, J. L., Volonteri, M., et al. 2021, MNRAS, 500, 3944
Rauch, K. P. 1999, ApJ, 514, 725
Regan, J. A., & Haehnelt, M. G. 2009, MNRAS, 396, 343
Reinoso, B., Schleicher, D. R. G., Fellhauer, M., et al. 2018, A&A, 614, A14
Rees, M. J. 1988, Natur, 333, 523
Rouan, D., Grosset, L., & Gratadour, D. 2019, A&A, 625, A100
Sakurai, Y., Yoshida, N., & Fujii, M. S. 2019, MNRAS, 484, 4665
Sanders, D. B., & Mirabel, I. F. 1996, ARA&A, 34, 749
Scharwächter, J., Combes, F., Salomé, P., et al. 2016, MNRAS, 457, 4272
Schleicher, D. R. G., Palla, F., Ferrara, A., et al. 2013, A&A, 558, A59
Shang, C., Bryan, G. L., & Haiman, Z. 2010, MNRAS, 402, 124
Shankar, F., Weinberg, D. H., & Miralda-Escudé, J. 2009, ApJ, 690, 20
Shlosman, I., Frank, J., & Begelman, M. C. 1989, Natur, 338, 45
Smajic, S., Moser, L., Eckart, A., et al. 2015, A&A, 583, A104
Soltan, A. 1982, MNRAS, 200, 115
Soria, R., Hau, G. K. T., & Pakull, M. W. 2013, ApJ, 768, L22
Storchi-Bergmann, T., & Schnorr-Müller, A. 2019, NatAs, 3, 48
Toomre, A. 1964, ApJ, 139, 1217
Toomre, A. 1981, Proceedings of the Advanced Study Institute (Cambridge: Cambridge Univ. Press), 111
Trujillo, I., Ferré-Mateu, A., Balcells, M., et al. 2014, ApJ, 780, L20
van den Bosch, R. C. E., Gebhardt, K., Gültekin, K., et al. 2012, Natur, 491, 729

van Velzen, S. 2018, ApJ, 852, 72
Venemans, B. P., Walter, F., Zschaechner, L., et al. 2016, ApJ, 816, 37
Vitral, E., & Mamon, G. A. 2021, A&A, 646, A63
Volonteri, M. 2010, A&ARv, 18, 279
Wang, R., Wagg, J., Carilli, C. L., et al. 2013, ApJ, 773, 44
Webb, N. A., Barret, D., Godet, O., et al. 2010, ApJ, 712, 107
Xiao, T., Barth, A. J., Greene, J. E., et al. 2011, ApJ, 739, 28

Chapter 3

AGN Feedback

The luminosity function of galaxies cannot be reproduced by the standard cosmological model ΛCDM if there is no active galactic nuclei (AGN) feedback, particularly necessary at high galaxy masses. There exist two main kinds of feedback processes, the radiative (quasar) mode, and the kinetic (radio) mode. The first is common for powerful AGN with high Eddington ratios, able to launch a fast wind from their accretion disk, the second is the low-luminosity ones, with mechanical action from their radio jets. The physical processes can distinguish the energy-driven and momentum-driven flows. Recent observations have revealed the large impact of molecular outflows, involving much more mass than ionized outflows. The best AGN feedback efficiency is detected in the center of galaxy clusters, where the brightest central galaxy (BCG) can regulate and stop the cooling flow of gas from the X-ray atmosphere.

3.1 Necessity of Feedback in Cosmology

Feedback processes have been introduced early in the development of galaxy formation theories (e.g., White & Rees 1978; White & Frenk 1991). In the standard cosmological model, ΛCDM, based on cold dark matter (CDM), where Λ represents dark energy (70% of the content), and the baryons represent no more than 5%, galaxies form in two steps. First the dark haloes form by dissipation-less collapse and become virialized, and then, just after recombination of the cosmic plasma into atomic hydrogen, the baryonic gas collapses inside the dark haloes, to form stars and galaxies. One of the main observations to reproduce is the luminosity function, i.e., the number of galaxies per unit volume as a function of their luminosity (cf Figure 3.1). Simulations of dark matter haloes predict a mass function, represented by the blue curve in Figure 3.1. The curve has been calibrated with a constant mass-to-light ratio of $M/L = 80$, to fit a range of luminosity around Milky Way-like galaxies. It appears clearly that in the dark haloes at the low and high mass ends, baryons have not been able to form as many stars, and/or some galaxies have not

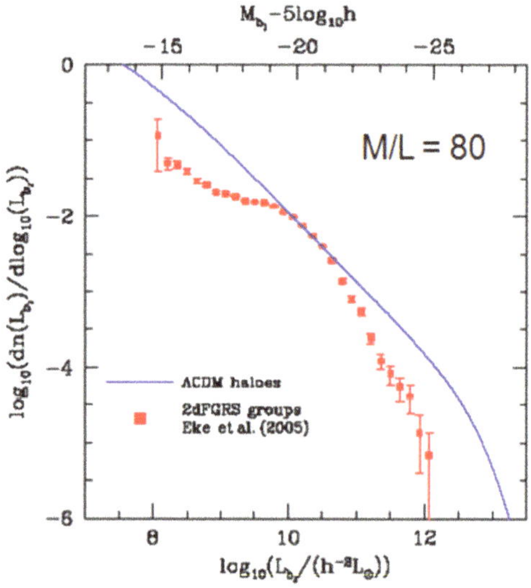

Figure 3.1. The observed luminosity function of galaxies (red squares with error bars, Eke et al. 2006) compared with the predictions of the ΛCDM simulations (blue line), assuming a mass-to-light ratio of $M/L = 80$. Simulations without feedback predict too many galaxies at small and large mass. Reprinted from Baugh (2006). © 2006. Reports on Progress in Physics. All rights reserved.

been able to form at all. From this observation, the stellar mass-to-halo mass ratio can be derived as a function of halo mass, and it shows a peak at 20% of the universal baryon to dark matter ratio at Milky Way-sized galaxies. This implies that more than 80% of baryons reside outside galaxies, which is interpreted by feedback mechanisms (e.g., Behrouzi et al. 2013).

At the low mass end, the supernovae feedback is able to solve the problem. The gravitational potential is smaller than the energy developed per unit mass by the supernovae, both through heating and mechanical processes, which is equivalent to an outflow velocity of ~100 km s^{-1}. For galaxies of circular velocity below this, supernovae can expel most of the gas from the potential (e.g., Baugh 2006). Also the primordial gas in a large fraction of the small halo population has been prevented to cool during the reionization epoch (e.g., Gnedin 2000). The background of high energy UV photons due to massive star formation in neighboring galaxies is able to photo-ionize their gas, and also to heat the intergalactic medium at a temperature higher than their virial temperature, so their potential is not deep enough to trigger gas infall.

As for the high mass end, these are not sufficient, and the AGN feedback is then required. Although the black hole has a very small mass (<0.5% of the bulge mass) and its sphere of influence, of radius $GM_{\rm BH}/\sigma^2$ is very small (σ is the bulge velocity dispersion), of the order of 20 pc for nearby Milky Way-like galaxies, its energy is sufficient to unbind gravitationally the bulge. During the growth of the black hole, accreting mass at a rate dM/dt and with a radiative efficiency of $\varepsilon \sim 0.1$, the luminosity of the AGN is $L = \varepsilon dM/dtc^2$, and the corresponding black hole growth is $dM_{\rm BH}/dt = (1 - \varepsilon)dM$ growth energies of the black hole and the bulge are in the ratio

$\varepsilon/(1 - \varepsilon)$ (M_{BH}/M_{bul}) $c^2/\sigma^2 \sim 400$. AGN have enough energy to unbind the whole galaxy. However, the coupling of this energy to the galaxy is in general very small, and the majority is radiated away, or a small fraction of the gas is expelled.

Also, the AGN feedback is significant only when gas accretion gives rise to a high luminosity. For low-luminosity AGN (low Eddington ratios), the thin accretion disk is reduced in extent, and the regime tends to ADAF (advection dominated accretion flow), with very low radiation. There is also a limitation for the most luminous quasars, which take their high luminosity from a thin accretion disk. The accretion disk in its outer parts is self-gravitating, and its stability can be described by the Toomre criterion. This relates the sound velocity c_s to the surface density and rotation frequency Ω of the disk. Since $c_s = H\Omega$, where H is the height of the accretion disk, there is a maximum radius of the accretion disk, beyond which the disk will fragment to form stars, and the supernovae feedback will expel the outer disk (King 2016). This results in a maximum radius for the efficient accretion disk, which is almost independent of the black hole mass, R_{disk} ~0.01 pc. On the inner side, the accretion disk is limited by the innermost stable circular orbit (ISCO), which is linearly dependent on the black hole mass. The ISCO radius is smaller for a prograde accretion than a retrograde one (cf Section 1.5.1 of Chapter 1). When the inner limit equals the outer self-gravitating limit, the accretion disk disappears, and there is no longer any luminous manifestation of the accretion. This phenomenon is the parallel for gas, of the maximum Hills limit of black hole mass, for which stars are swallowed whole without liberating any energy (cf Section 2.1 from Chapter 2). The resulting maximum black hole mass, as a function of the BH spin a (dimensionless quantity proportional to the angular momentum J) is displayed in Figure 3.2. The position of some of the most massive AGN (in red in the figure) sometimes put a constraint on the spin parameter a. The latter must be positive, and even $a > 0.2$ for the 0014 + 813 quasar.

3.2 Two Modes of Feedback

According to the luminosity class of AGN, two distinct modes of feedback may be distinguished, the radiative or quasar mode, and the kinetic or radio mode. When the quasar luminosity is very high, near the Eddington limit, the AGN is powered by a thin accretion disk. Then the heating and the radiation pressure are able to evaporate a wind from the accretion disk. This is the quasar mode (wind). The Eddington's luminosity is reached when the radiation pressure compensates for gravity. The radiation pressure due to the absorption of photons by ionized hydrogen, is $\kappa/c\ L/(4\pi r^2)$ where κ the opacity is here $\sigma T/m_p$, σT is the Thomson cross section and m_p the mass of the proton. Since the gravity force is GM/r^2, the radius dependence cancels out, and the Eddington luminosity is $L_{Edd} = 4\pi c G M m_p/\sigma T$. The ionized gas can no longer flow freely and part of it escapes the galaxy in a high velocity wind. The limiting luminosity can be estimated as $L_{Edd} = 3.3 \times 10^{13} L_\odot$ $(M/10^9 M_\odot)$. This limit is valid for a spherical gas distribution, but may be bypassed, if the material is distributed on a disk. The accretion rate corresponding to this limit is $20 M_\odot$/year $(M/10^9 M_\odot)$. This limit is relevant for the ionized gas disk, since it

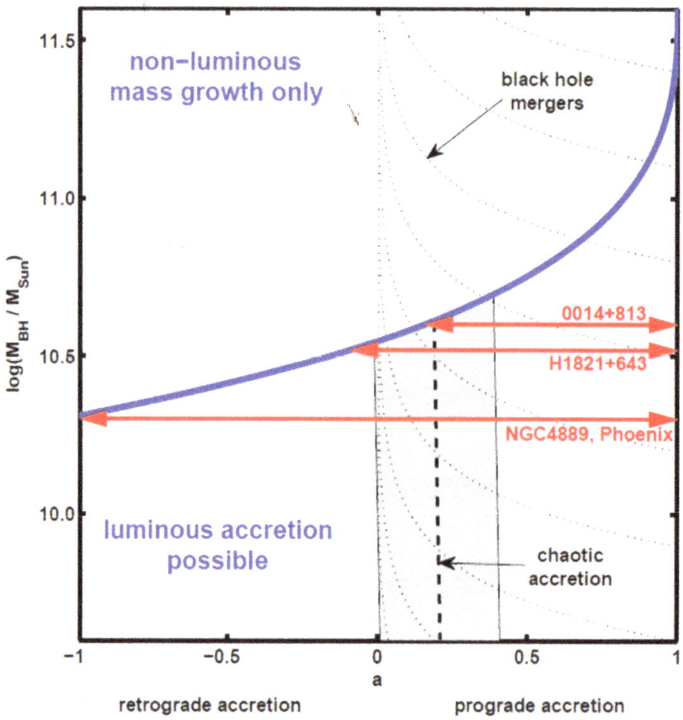

Figure 3.2. The maximum black hole mass (blue curve) for a quasar, as a function of the spin parameter $a = cJ/(GM_{\rm BH}^2)$ (where J is the BH angular momentum). Negative (positive) values of a correspond to retrograde (prograde) accretion. Visible quasars or AGN lie below the curve. Three of the most massive AGN are plotted in red: 0014 + 813, with $M = 4 \times 10^{10}\ M_\odot$ (Ghisellini et al. 2010), the central quasar of the H1821 + 643 cluster ($M = 3 \times 10^{10}\ M_\odot$, Walker et al. 2014), NGC 4889 (McConnell et al. 2011), and the central galaxy of the Phoenix cluster (McDonald et al. 2012), both having $M = 2 \times 10^{10}\ M_\odot$. The dotted curves show how the spin of black holes statistically decrease in mergers (Hughes & Blandford 2003), i.e., high masses have in average lower spins. The vertical black dashed line, with its gray spread, represents the average angular momentum acquired after several randomly oriented accretion episodes, for chaotic gas accretion (King et al. 2008). Reproduced from King (2016).

involves the Thomson cross section σT for scattering between photons and charged particles. There is also another limit, which may involve gas further out, provided that if contains dust; the limit is derived from the dust cross section σd and the ratio between the two cross section is $\sigma d/\sigma T = 1000$ for a dust-to-gas ratio typical of the Galactic interstellar medium. The Eddington limit is proportional to the mass of the illuminated object, and inversely to the cross section with the photons. With the dust cross section, the same $L_{\rm Edd}$ involves now an object 1000 times more massive, which corresponds to the bulge of the host galaxy, according to the scaling law. Dust and gas are coupled through collisions and the dust can therefore entrain the gas out. This might favor the building of the scaling relation (Fabian 2012). To have an efficient feedback effect, the quasar mode requires high densities of relatively cold gas around the black hole, for instance large-scale inflows triggered by galaxy mergers. When the coupling is not sufficient, the feedback has negligible impact on the galaxy (e.g., Gabor & Bournaud 2014).

For the low-luminosity AGN, below 10^{-2} Eddington, the possible feedback mechanism are rather mechanical, by the pressure and dragging of the relativistic jet, and by the shocks and subsequent heating of the gas in a bubble and cocoon inflated by the jet. This is the radio mode. Radio jets are only observed in about 10% of all AGNs. The radio mode is frequently observed for the central radio galaxies in clusters (BCG), where they can stop or moderate cluster cooling flows. The black holes significantly influence the evolution of their host galaxies, reducing their amount of cold gas and quenching their star formation. In addition, the properties of intracluster gas are affected strongly by the presence of massive black holes in the cores of galaxy clusters, leading to shallower metallicity and entropy profiles.

The presence of powerful and long radio jets (quasars called "radio loud") is statistically anti-correlated with the brightness of the quasar. More precisely, the ratio between radio and optical brightness $R = L_{rad}/LB$ is a decreasing function of the Eddington ratio λ. The low-luminosity AGN, corresponding to low λ, lower than 10^{-3}, are frequently in the ADAF regime, with a reduced size and a large height of the disk. In this regime, relativistic MHD simulations have shown that both relativistic jet and a disk wind can provide some feedback (see Figure 3.3). The jct has much morc cncrgy than thc wind, but a wcak coupling with thc circum-nuclear gas. Although much weaker, the wind provides efficient feedback on a wide range of scales, since it subtends a large solid angle in the host galaxy (Sadowski et al. 2013).

3.3 Energy or Momentum Conservation

The physical processes able to provide energy and momentum to a gas wind or outflow are numerous (e.g., Veilleux et al. 2020 for a review): first the AGN radiation pressure, as described above, especially when close to the Eddington luminosity, the relativistic radio jet and its ram-pressure, but also all subsequent processes, as thermal pressure after heating of the gas, the cosmic ray pressure, when shocks have been produced, and cosmic rays are accelerated at non-thermal energies by diffusive shock acceleration (DSA), as in supernovae. Of course, starbursts could be triggered as often when the central BH is fueled, and the supernovae processes may add in concert to provide more energy. Taking into account of all these processes would require sophisticated simulations, but simple considerations may however be made, to distinguish the different regimes of the flows. In the simplification of spherical geometry, it is possible to describe the characteristic shock waves, as well known in supernova bubble models. Figure 3.4 (right) shows schematically the different spheres, reverse shock wave, surface of discontinuity between wind and interstellar medium, and shock wave advancing in the galaxy. The wind is launched in the galactic nucleus with the initial velocity V_{in}, which may be up to 10% the light speed, as seen through X-ray velocities (e.g., Tombesi et al. 2010);

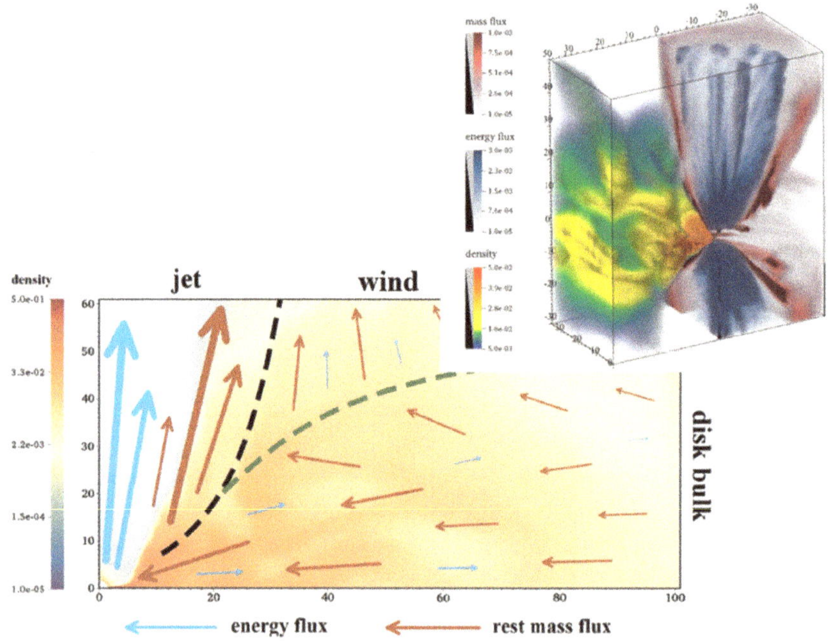

Figure 3.3. Bottom: schematic drawing of the various components around a spinning black hole: the relativistic jet in the center, the wind at the top of the disk, above the green-dashed line, and the rotating disk. The simulation at the top right shows in blue the energy flux, in purple, the mass flux and in green–yellow–red, the density. Reproduced from Sadowski et al. (2013).

after the shock at R_{sw}, it enters the shocked-wind zone (b), where possible radiation could occur. While adiabatic expansion (energy-conserving) was first assumed by Silk & Rees (1998), others argued that the gas was cooled by inverse Compton from the quasar radiation (e.g., King 2003; Reeves & Braito 2019), and therefore the flow is only momentum-conserving. A second shock, at radius R_s (velocity V_s), is driven into the ambient interstellar medium. The two media are separated by a contact discontinuity at R_c. It is the cooling of the shocked-wind, zone (b) in Figure 3.4 (not the shocked ambient medium, zone (c)) that determines whether the outflow conserves energy or momentum.

The momentum flux carried by the wind $\dot{M}\mathrm{w}\, V_{in}$ is equal to $\sim L_{AGN}/c$, or $\tau L_{AGN}/c$ in case of dust with opacity $\tau < 1$, and the energy of the wind ½ $\dot{M}\mathrm{w}\ V_{in}^2$. The outflow after the shock at shell radius R_s has energy ½ $\dot{M}\mathrm{s}\ V_s^2$. The energy-conserving condition can be written as $\dot{M}\mathrm{s}\ V_s^2 = f\dot{M}\mathrm{w}\, V_{in}^2$, with $f < 1$ being the efficiency of conservation. It is possible for instance to assume that some fraction of the energy is spent to inflate the bubble. The momentum of the outflow $\dot{M}\mathrm{s}\ V_s = \dot{M}\mathrm{w}\, V_{in}\ (fV_{in}/V_s)$ is boosted from the AGN momentum L_{AGN}/c by the factor fV_{in}/V_s, which can be as high as 50 (Faucher-Giguère & Quataert 2012). This explains the observed statistics of molecular outflows in AGN galaxies, where the momentum flux is in average 20 times L_{AGN}/c (Cicone et al. 2014).

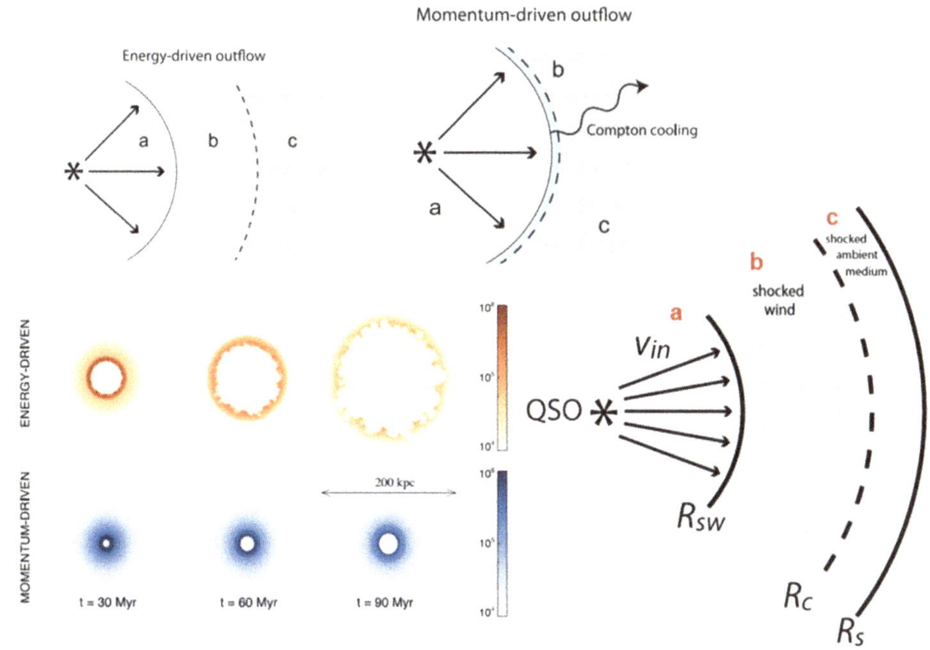

Figure 3.4. Diagram (right) of the various outflow phases around an AGN in the radiative phase (quasar mode). R_c is the discontinuity surface, R_s the shock in the interstellar medium, R_{sw} the reverse shock towards the relativistic wind. Reproduced from Faucher-Giguère & Quataert (2012). © 2012. The Royal Astronomical Society. All rights reserved. The zone "a" is characterized by the initial speed V_{in}, then the zone "b" of the shocked wind, which can be considerably reduced in volume, if the cooling is effective, as in the hypothesis of a momentum-conserving flow, which does not conserve energy. The "b" zone remains important, in the case where the energy is conserved, as shown by the simulations of Costa et al. (2014). In the middle, in orange, the gas shell extends very far radially, in the case where energy is conserved. At the bottom, in blue, the extension of the gas is much smaller, because energy is not conserved. Each panel corresponds to a 200 kpc square, for a black hole of mass $10^8 M_\odot$ which emits the Eddington luminosity. The stages of the simulation are 30, 60 and 90 million years. Reproduced from Costa et al. (2014). © 2014. The Royal Astronomical Society. All rights reserved.

In summary, in the energy-conserving regime, most of the energy of the AGN is not radiated away, but serves to push the ambient medium, to produce work and transfer momentum to the gas. Simulations confirm how strongly the efficiency of the feedback depends on the regime of the flow: if the energy is conserved in the shocked-wind region, the shell in the ambient medium will be more extended, and the impact on the interstellar medium much larger than in the momentum-conserving regime (Costa et al. 2014). This implies that the black hole growth is only significant in the momentum-driven flows, while in energy-driven ones, the AGN will expel most of its fueling gas. It is important then to determine the conditions of these phases.

For that, it is paramount to compare the flow timescale to the cooling timescale. The inverse Compton cooling is due to the AGN radiation, and if the wind shock is close to the nucleus, the radiation field is so intense that the cooling time is shorter than the flow time R_{sw}/σ. The radiation energy density decreases as R^{-2}, thus the cooling time increases as R^2. The flow time increases only as R, so there is a critical radius, depending on $M_{\rm BH}$ and the velocity dispersion σ, that separates the two regimes. This critical radius is of the order of 1 kpc for nearby active nuclei (King et al. 2011). The regime is momentum-driven inside ~1 kpc, and energy-driven outside.

The simplification of spherical symmetry may be used even for shells extending over only a fraction of the 4π solid angle. There is however another important factor, the timescale of AGN activity, inside their duty-cycle. Variability is frequently observed over years, months and days, especially close to the nucleus, in the region of fast outflow in UV and X-rays (Tombesi et al. 2010). Taking this variability into account, and noting that black holes grow their mass essentially in momentum-driven phases, it is quite tantalizing to imagine various phases in the black hole evolution (e.g., King & Pounds 2015). At the beginning, the black hole is under-massive, and its mass is below the M–σ scaling relation. The fast AGN wind is quickly stopped by a shock, at scale much lower than 1 kpc and energy is radiated away in the shocked-wind zone. In this momentum-driven regime, the mass of the black hole grows steadily. Then the mass of the black hole reaches the M–σ scaling relation, the flow is now extended beyond kpc scale, and enters the energy-driven regime. The pressure from AGN activity is much larger than the local pressure (ρv^2) of the ambient medium, and the ambient gas is easily ejected. The cycling of these two phases is providing a regulation of the mass of the bulge. In this scenario, the galaxy disk is a massive obstacle, that deviates the wind, which becomes then bipolar.

3.4 Molecular and Ionized Gas Outflows

3.4.1 Nearby AGN Galaxies

Ionized gas outflows have been observed for a long time, through blue-shifted absorption lines in the UV and X-ray spectra of luminous AGN (Weymann et al. 1991; Tombesi et al. 2010). The most extreme winds are observed through broad absorption lines, in BAL QSO, with broad C IV absorption lines of width larger than 2000 km s^{-1}, or the more frequent mini-BALs of widths between 1000–2000 km s^{-1}. At least half of local AGN show the presence of low-velocity narrow absorption lines with blueshift of a few 100–1000 km s^{-1}, both in the UV and in the X-ray band (e.g., the review by Crenshaw et al. 2003). Coronal forbidden lines are observed extended with blue-shifted asymmetries in nearby Seyferts, implying very ionized gas outflows with velocities up to 2000 km s^{-1} (Rodriguez-Ardila et al. 2006). At a larger distance from the nucleus, neutral gas outflows have been observed, either atomic or molecular. Atomic gas may be observed in radio through the HI-21 cm line (Morganti & Oosterloo 2018), and also the Lyα line in the UV, and atomic sodium doublet NaI D in absorption in the optical spectrum (Veilleux et al. 2020).

Molecular outflows have been observed first with the CO molecule, through large wings in the central spectrum, corresponding to outflow velocities of ~1000 km s^{-1} (Figure 3.5, e.g., Feruglio et al. 2010). The first and prototypical case was Mrk231, both a powerful quasar and a starburst (ultra-luminous IR galaxy, ULIRG).

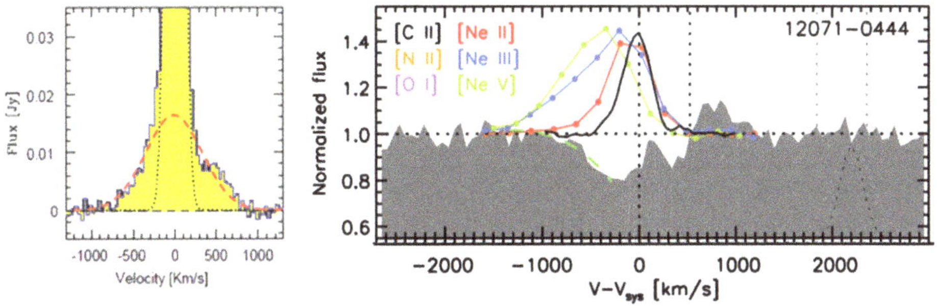

Figure 3.5. Left: CO spectrum of Mrk231, the prototypical quasar and starburst (ULIRG), made with the IRAM-30 m telescope. Reproduced with permission from Feruglio et al. (2010). © ESO. The spectrum is fitted with a central Gaussian in black-dotted line, and wide wings with red-dashed line. Right: a Herschel/PACS spectrum of the 119 μm OH line (gray shading) superposed with mid-IR and far-IR fine structure lines of [C II], [Ne II], [Ne III], [Ne V] in black, red, blue and green respectively, observed with Spitzer. The OH absorption is fitted with a green-dashed line. Reprinted from Spoon et al. (2013). © 2013. The American Astronomical Society. All rights reserved.

Mrk231 is a BAL system, showing outflows in the ultraviolet and optical domain, with a blue-shifted wing extending up to 7000 km s^{-1} (Veilleux et al. 2016). Starbursts and AGN are frequently associated, and both supernovae and AGN feedback are playing a role, to enhance the gas outflow, which may extend up to kpc size.

Molecular outflows have also been observed in the far-infrared (119 μm OH line, cf Figure 3.5) with Herschel/PACS in dozens of local galaxy mergers, mostly ULIRGs and quasars (Veilleux et al. 2013). The OH line reveals a neat preference for blue-shifted absorption: in 70% of cases velocities bluer than 50 km s^{-1} are detected in absorption, while less than 10% are detected in redshifted absorption. These statistics imply a wide opening of the outflow, about 145°, while the inflow might be in a plane. Typical outflow velocities are 200 km s^{-1}, some reaching 1000 km s^{-1}. The presence of strong outflow is correlated to the presence and strength of the AGN.

One of the first statistical studies about molecular outflows in nearby galaxies by Cicone et al. (2014) has revealed two class of objects: the starburst dominated and AGN dominated galaxies. Only in the latter, the mass outflow rate dM/dt, computed assuming spherical or multi-conical geometry, $dM/dt = 3\ V_o\ (M_o/R_o)$ with M_o, R_o, V_o being the mass, radius, and velocity of the measured outflow, is proportional to the bolometric luminosity of the AGN. Also the moment flux VdM/dt is correlated to the AGN luminosity and is equal to 20 L_{AGN}/c. This suggests an energy-driven molecular outflow (cf Section 3.3).

The outflows have a multiphase character, however the mass outflow rate of the ionized gas is much lower than that of molecular gas. This is important for the feedback impact, only the loss of molecular gas is able to moderate the star formation rate of the galaxy. If the outflow velocity does not reach the escape velocity, the star formation will be only postponed. The study of scaling laws between mass flow rate and AGN luminosity has revealed that the slope of the power-law for the molecular gas is lower than for the ionized gas (cf Figure 3.6, Fiore et al. 2017). The mass flow rate is much higher (by 1–2 orders of magnitude) in

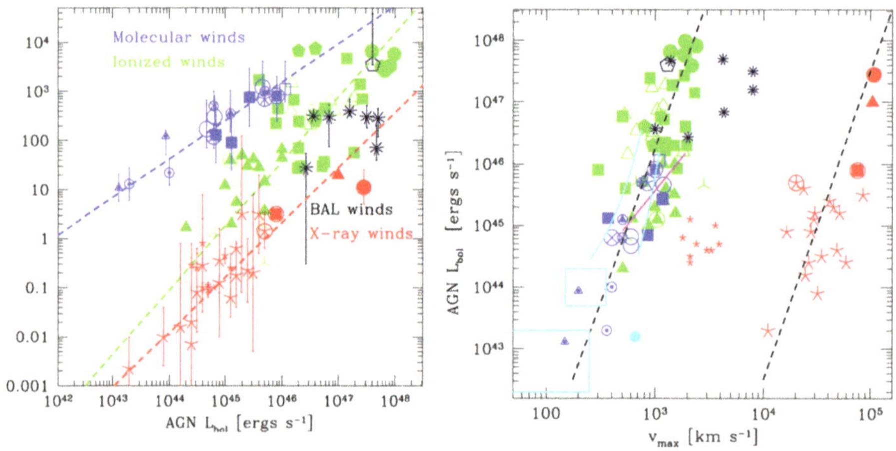

Figure 3.6. Left: mass outflow rate as a function of AGN bolometric luminosity, for molecular gas (blue), ionized gas (green), X-ray outflows (red), and broad absorption lines (BAL) in black. Right: AGN bolometric luminosity as a function of the maximum wind velocity, V_{max}. The dashed black lines trace power-laws of slope 5. Reproduced with permission from Fiore et al. (2017). © ESO.

molecular than ionized gas at low AGN luminosity, but the difference reduces with luminosity, and both rates tend to a common value at the most extreme luminosities. On average, the flux of kinetic energy in the gas outflows is 2.5% of the bolometric luminosity. From these mass loss rates, it is possible to derive a molecular gas depletion timescale, which is 3–10 times shorter, for the most luminous AGN, than the depletion timescale of main sequence galaxies with similar star formation rate and stellar mass. This proves the efficiency of feedback for the powerful AGN, but not for the low-luminosity end. The right panel of Figure 3.6 shows a very good correlation between the AGN luminosity and the maximum velocity of the outflow (power-law of slope 5), for both ionized and molecular gas. For the X-ray flow, there exist two groups: the warm absorbers, with a moderate velocity (~2000 km s^{-1}), and the ultra-fast outflows (UFO) with $V = 10^4$–10^5 km s^{-1}. A recent study with larger less-biased samples, reports more scatter in the energy flux, comprised between 0.1 and 5% of the AGN luminosity, and in the momentum flux, between 1 to 30 L_{AGN}/c (Fluetsch et al. 2019), suggesting a large range between momentum-driven and energy-driven outflows. With larger statistics, it is possible to see that about 10% of outflows have energetics that exceed the maximum theoretical values, based on the observed L_{AGN}. This is interpreted as time variability, the observed outflows being a witness of past higher activity of the nucleus (Fluetsch et al. 2019). In less than 5% of cases, the gas has enough energy to escape the galaxy, implying a very moderate feedback efficiency, limited to the central regions.

A very instructive example of outflow is the prototype of Seyfert-2 galaxy, NGC 1068. This is one of the nearest powerful AGN, its Eddington ratio is 0.5, for a black hole mass of $1.6 \times 10^7\ M_\odot$. It is located at a distance $D = 14$ Mpc (1 arcsec= 70 pc), has a primary bar ($R = 8$ kpc, PA = 5°) with a star-forming pseudo-ring at Lindblad resonance of radius 1 kpc. Inside the ring is embedded a nuclear bar (PA = 48°) and well inside a circumnuclear disk (CND) of radius ~150 pc (Schinnerer et al. 2000). NGC1068 has a radio jet starting perpendicular to the nuclear water maser disk (Gallimore et al. 2004) and is deflected by molecular clouds in the disk, at 100 pc scales. It is clear that the accretion disk and maser disk have not the same orientation as the main disk; the latter is almost face-on, while the former are edge-on. Given the high Eddington ratio and the radio jet sweeping the galaxy plane, NGC1068 is a galaxy where both radiative mode and radio mode of feedback may play a comparable role, as shown in Figure 3.3.

The hierarchical structure of embedded bars favors the AGN feeding. The various structures are well identified in the molecular component, traced by various

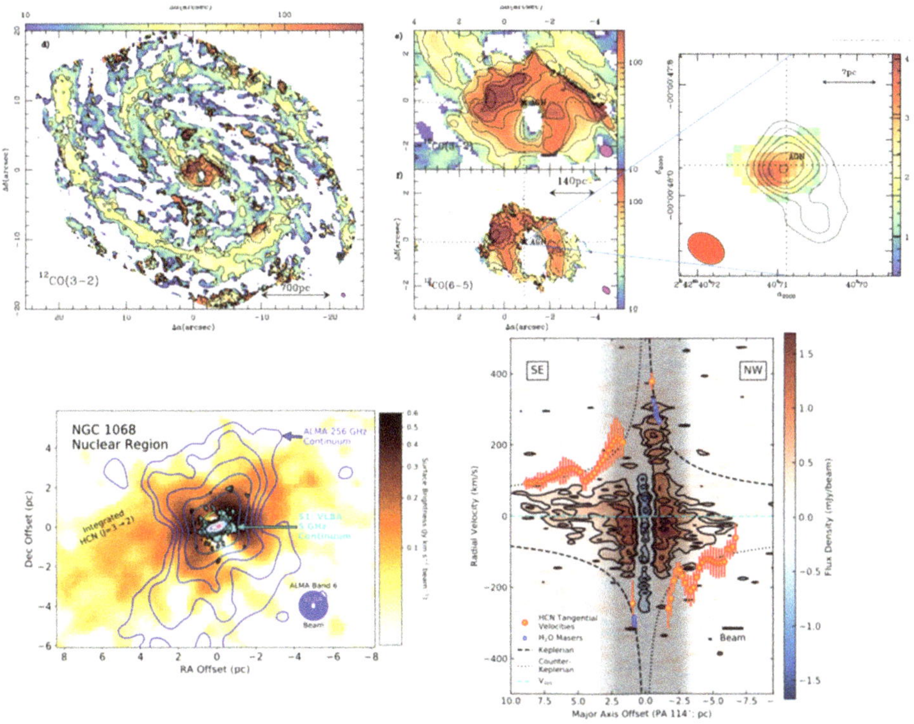

Figure 3.7. Top left: CO(3–2) ALMA map of NGC 1068, with a field of view of 50″ in right-ascension, flanked at right with a central zoom, with 9″ in right-ascension (CO(3–2) top, and CO(6–5) bottom), from Garcia-Burillo et al. (2014, 2016). At the extreme right, the zoom is on the AGN position in CO(6–5), the field of view is now 0.4″ on each side of the box, from Garcia-Burillo et al. (2016). The superposed contours are the continuum at 685 GHz. Reproduced with permission from Garcia-Burillo et al. (2014, 2016). © ESO. Bottom left: ALMA map of HCN(3–2) in colors, with the 256 GHz continuum contours superposed in blue, and VLBA 5 GHz continuum contours in black, filled with shades of cyan to magenta, from Impellizzeri et al. (2019). Bottom right: Major-axis position–velocity diagram for HCN in the nucleus, showing the counter-rotation The tangential HCN velocities are marked by orange dots with red error bars, and the H_2O maser radial velocities as filled blue circles. The Keplerian rotation corresponding to the masers is indicated by the dashed line (and in the opposite side by dotted lines). The gray-shaded region represents the extend of the 256 GHz continuum emission. Reprinted from Impellizzeri et al. (2019). © 2019. The American Astronomical Society. All rights reserved.

molecules, at increasing resolution, as displayed in Figure 3.7. Inside the CND, the AGN is off-centered, and is surrounded by a molecular torus, of radius 3–7 pc, depending on the tracer (Garcia-Burillo et al. 2016, 2019). The dust emission at 0.4 mm wavelength shows an elongation along the polar direction, parallel to the synchrotron emission of the radio jet (Figure 3.7). This is also the orientation of the near infrared dust emission, following a hollow cone (Figure 3.8), well delineated in polarization (Gratadour et al. 2015). At sub-pc scale, the Gravity Collaboration (2020) has seen a ring with the same orientation as the maser disk, which could be the sublimation radius of the dust. has The ionized gas outflow has been known for a long time (e.g., Crenshaw & Kraemer 2000; Vermot et al. 2019), and is compatible with a biconical outflow, with an orientation similar to the radio jet, and a velocity of 300–1000 km s^{-1}. The molecular gas outflow is following the same orientation, with a mass outflow rate of 63 $M_{\odot}$ yr^{-1} (Garcia-Burillo et al. 2014). Curiously, the rotation velocity is quite low in the CND and torus, but this may be explained by counter-rotation (Impellizzeri et al. 2019; Imanishi et al. 2020). This phenomenon may be attributed to successive cloud accretion onto the CND with random angular momentum.

There is some evidence of several origins for the molecular outflow: the gas is expanding both in the equatorial direction of the torus, and in the perpendicular direction, along the biconical outflow delineated by dust. The molecular torus has a

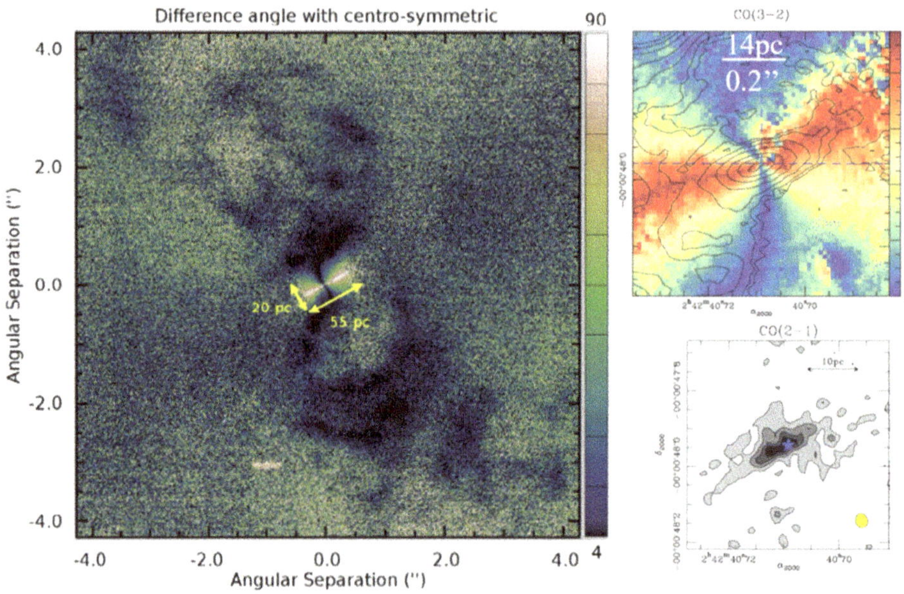

Figure 3.8. Left panel: adaptive-optics polarimetric image of NGC1068 in the near infrared, with SPHERE at the VLT. The image is the result of the difference between the polarization angle map and a purely centro-symmetric pattern. Reproduced with permission from Gratadour et al. (2015). © ESO. Right panel: at top, the CO(3–2) ALMA contours are superposed on the central zoom of the left image; bottom, the CO(2–1) at the same scale. Reproduced with permission from Garcia-Burillo et al. (2019). © ESO.

stratified structure, smaller in the high-density tracers (HCN, HCO+) than in the CO tracer. The CND is empty in the center, and its inner surface is corresponding nicely to the near infrared polarized emission, that can be identified to the working surface of the ionized wind. A wind from the accretion disk is therefore pushing the molecular gas over a wide-angle, together with a radio jet, entraining and shocking the gas in a narrower solid angle (Garcia-Burillo et al. 2019).

Another instructive case, which is now from the low-luminosity regime, is NGC 1377. This is a nearby (D = 21 Mpc, 1″ = 100 pc) early-type galaxy (lenticular), which is extremely radio-quiet, i.e., its radio to infrared luminosity ratio is well below the normal sequence for star-forming galaxies (Helou et al. 1985). There is a radio source in the center, detected in synchrotron emission with the VLA, but no radio jet detected. Nevertheless, it is the most collimated molecular outflow yet detected (Aalto et al. 2016, 2020). First ALMA CO(3–2) observations with moderate resolution (0.2″) revealed an extremely collimated, 150 pc long molecular jet (Figure 3.9, in red and blue), together with a wide-angle wind flow, in gray (with smaller velocities). No detected outflow in ionized or atomic gas has been reported. The molecular outflow appears to be almost in the plane of the sky, so well aligned, as a slight precession of the jet is observed: there is a reversal of the blue-shifted emission to redshifted in the north, and the reverse in the south (Figure 3.9). This has been interpreted with a precession around the average angular momentum axis by an angle of 10°–15° (Aalto et al. 2016). Given the low inclination with rcspcct to the

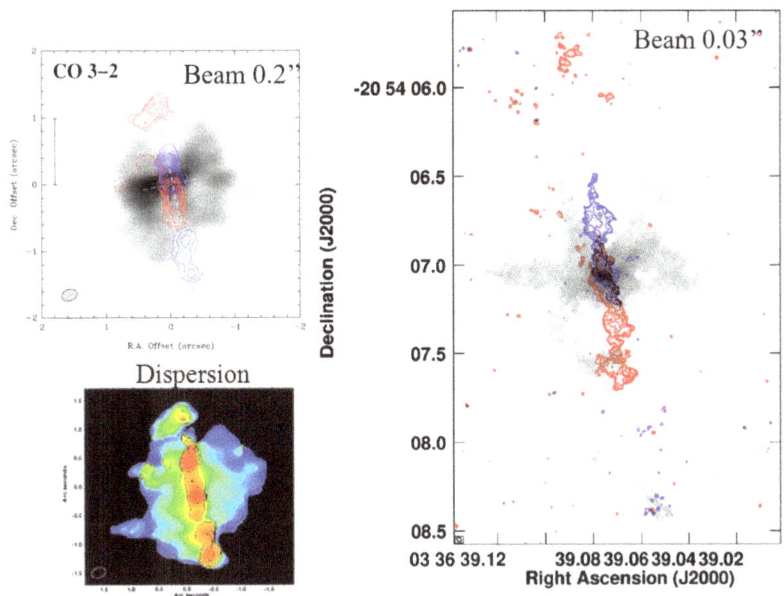

Figure 3.9. Top left: CO (3–2) integrated intensity image where emission close to systemic velocity (0–70 km s^{-1}) is shown in grayscale. The high-velocity wings (80 to 160 km s^{-1}) from the molecular jet are shown in red and blue contours (for the redshifted and blue-shifted gas respectively). Right: same at high resolution (beam 0.03 × 0.022″). Bottom left: the corresponding velocity resolution, at low resolution, showing clearly the molecular jet. Reproduced with permission from Aalto et al. (2016, 2020). © ESO.

sky plane, the outflow velocity is uncertain, between 200 and 800 km s^{-1}. The mass in the jet is estimated at 2×10^6 $M_\odot$. Higher resolution observations have confirmed that the molecular outflow is very well collimated to the spine of the jet (Figure 3.9). This cannot be done alone, with a neutral gas, through star formation, and should be due to the magnetic collimation of a ionized radio jet (Aalto et al. 2020). The molecular jet is therefore the signature of a past radio jet and past radio activity of the AGN.

In the nearby universe, powerful AGN are rare, the large majority have Eddington ratios below 0.01. Most of the molecular outflows are therefore dominated by the radio mode. The coupling is however excellent, since the orientation of the accretion disk is observed to be randomly distributed with respect to the galaxy disk (e.g., Schmitt & Kinney 2002). Frequently, the radio jet encounters molecular clouds in the host galaxy disk (e.g. NGC 4258; Cecil et al. 2000). This is particularly the case for IC5063 (Morganti et al. 2013a; Dasyra et al. 2016) and 4C12.50 (Dasyra et al. 2012; Morganti et al. 2013b). In the case of IC5063, the radio jet is launched to run exactly in the disk plane, and produces shocks and a jet bubble, implying a strong ionized, atomic and molecular gas outflow. 3D relativistic hydrodynamic simulations succeed to reproduce many characteristics of the observations, including the high turbulence in the plane, the deflection of the jet plasma encountering molecular clumps, and the wind propagation perpendicular to the plane (Mukherjee et al. 2018). At the zone of impact with the clouds, the jet splits in several streams, and the radio emission is maximum. The cooling timescale is still relatively long with respect to the dynamical timescale, and molecular clouds are deflected and displaced in block, rather than re-formed *in situ*.

The discovery of massive molecular outflows, composed of dense molecular clouds, was a surprise. How can molecular clouds survive the high velocity ~1000 km s^{-1} shocks, able to dissociate molecules? What are the processes able to entrain and accelerate molecular clouds in a radio jet? This might happen in several steps. Molecular clouds are embedded in a more diffuse warm atomic medium and then a hot and diffuse ionized gas. The AGN strong wind can first create a flow in the diffuse phases of the ambient medium. A relativistic radio jet also creates shocks at the interface with this medium, resulting on a weaker and slowed down wind, creating pressure on the multiphase medium and on the dense molecular clouds. The internal pressure gradients can shred and disrupt the clouds, which will gain a higher surface, and be more sensitive to the flow. Their cross section will be enhanced, their density decreased, they will be more coupled to heating and ionization, instabilities. According to the wind strength, it is possible that entire molecular clouds are evaporated, while they acquire momentum from the wind. However, after the shocks, instabilities will create clumps again, and radiative cooling will form anew the population of molecular clouds (Figure 3.10). This process on a multiphase medium has been simulated and found to be quite efficient (Hopkins & Elvis 2010). Realistic clouds have been taken into account, and the time-dependent chemistry of 157 species, of which 20 molecules, to show that the cooling of the evaporated cloud and its clump fragmentation could occur in typically 1 Myr (Richings & Faucher-

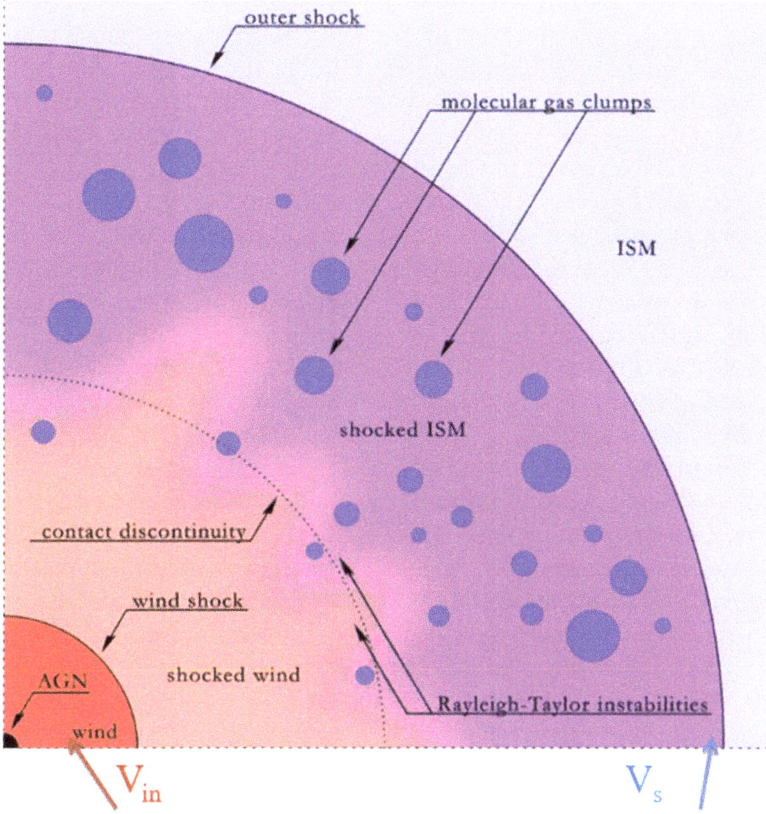

Figure 3.10. Schematic spherically symmetric drawing of the energy-driven outflow, when the AGN is a high luminosity quasar, able to drive an almost relativistic wind (V_{in} ~0.1c) from its accretion disk. The wind is first shocked strongly closed to the disk, and after a contact discontinuity, it expands and drives an outer shock into the host interstellar medium (ISM). The contact discontinuity is Rayleigh–Taylor unstable, the wind and ambient medium mix there, and molecular clouds can form *in situ*. The outflow cools radiatively, and can fragment in cold molecular clumps. Despite its high velocity, up to $V_s = 1000$ km s^{-1}, most of the outflow is in molecular form. Reproduced from Zubovas & King (2014).

Giguère 2018a). Escape velocity can be reached, and the characteristics of the observed molecular outflows can be reproduced, provided a minimum density and metallicity (and dust) content. Dust grains are also able to re-form after the shocked ambient gas has cooled, provided that a small fraction of initial dust grains have mixed with the swept up and cooled layer and serve to catalyze rapid molecule formation (Richings & Faucher-Giguère 2018b)

3.4.2 Central Galaxies in Clusters

Galaxy clusters containing a cool core are among the best examples of efficient AGN feedback. From massive clusters down to rich groups of galaxies, these large-scale dark matter haloes contain a very hot gas, emitting X-ray radiation. The temperature of this gas can reach up to 10^8 degrees for the richest clusters, in Virial equilibrium in the potential well of the cluster. The average cooling timescale of this gas over its volume is greater than the Hubble time. Some clusters have their temperature decreasing clearly towards the center, they are the cool core clusters. Their X-ray radiation has a remarkable peak at the center. Since the cooling rate is proportional to the square of the density, the cooling time t_{cool} decreases towards the cluster center and becomes less than the Hubble time, typically $t_{\text{cool}} \sim 300$ million years. In the absence of feedback, the gas would therefore cool, lose its pressure support and fall towards the central galaxy. The latter called brightest cluster galaxy (BCG) is a very massive elliptical galaxy, which has swallowed many companions, and satellite galaxies driven towards the center by dynamic friction. The majority of them are host of a radio-loud AGN (Best et al. 2007). In a very simple view of spherical symmetry, and no feedback, the cooling rate should imply a gas accretion onto the BCG up to 1000 $M_{\odot}$ yr^{-1}, which is not observed. The young stars that this cold gas should have formed are not detected either. It is now generally accepted that some form of heating balances the radiative cooling, leading to cooling flow rates an order of magnitude lower, on the order of 10 $M_{\odot}$ yr^{-1} (e.g. Peterson & Fabian 2006).

Multi-wavelength observations soon convinced astronomers that mechanical and radiative heating from the central AGN was the source of this feedback. X-ray detailed images by Chandra revealed the complex structure of the intracluster medium (ICM), in particular bubbles, or cavities, symmetrically distributed on each side of the center, one of the prototype being the nearby cool-core Perseus cluster (cf Figure 3.11). Radio observations of the jets revealed a close correlation with the bubbles, demonstrating that the cavities are sculpted by the interaction of the radio jets and the ICM gas. Each cavity is surrounded by a bright and high-pressure region in X-ray, and shocks beyond, with frequently cold fronts (e.g. the review by Fabian 2012). Ionized gas filaments, most of the times directed towards the center, are radiating in the Hα line (cf Figure 3.11). These filaments in Perseus are up to 50 kpc long, essentially falling onto the central BCG, sometimes with a horse-shoe shape to flow around a cold front (Conselice et al. 2001; Fabian et al. 2003). These Hα filaments are not excited by star formation, nor by the AGN (since the excitation should decrease with distance). They reveal that a multiphase gas can coexist within the hot atmosphere, here with a 10^4 K phase. These filaments are also delineated by the ro-vibrational 2 μm emission of H_2 molecules, tracing shocks (Lim et al. 2012). Their excitation could be mainly due to cosmic rays (Ferland et al. 2009).

The gas is cooling down to a molecular phase at $T \sim 10$ K. In the last two decades, astronomers have detected cold molecular gas at the center of the cool core clusters, corresponding to 10% of the expected unlimited cooling rate, and understood that this flow of gas was at the origin of the nuclear activity of the BCG. The latter is able then to moderate the cooling rate, by the associated feedback phenomena (Edge 2001; Salomé & Combes 2003; Olivares et al. 2019; Russell et al. 2019).

The feedback process in cool core clusters is based on the radio mode. The central BCG are low-luminosity AGN, radio loud, in the FRI class, as the Perseus A source. The relativistic radio jet propagates into the ICM, creating a cavity. The gas inside the cavity is diffuse and hot, and in approximate pressure equilibrium with the

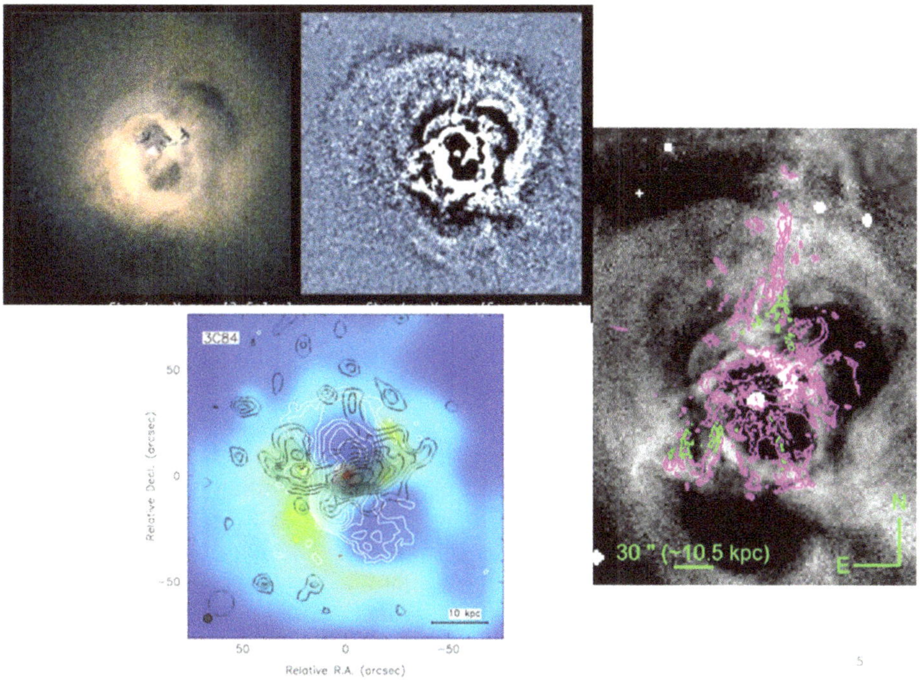

Figure 3.11. Multi-wavelength images of the cool-core Perseus cluster. At top left is the X-ray Chandra image, showing cavities, and at its right side, the unsharp-masked image, revealing the ripples. Reproduced from Fabian et al. (2003). © 2003. The Royal Astronomical Society. All rights reserved. In the bottom left, the X-ray image is reproduced in colors, with the VLA radio image in white contours, and the molecular gas (CO(2-1) emission) in black contours. Reproduced with permission from Salomé et al. (2006). © ESO. At right the unsharp-masked image is superposed by the ionized gas in pink contours, and young star formation in green. Reproduced from Canning et al. (2014). © 2014. The Royal Astronomical Society. All rights reserved.

outside ICM. Since it has a lighter density, the bubble is buoyantly rising upward. A typical activity period of the AGN and radio jet is 40 Myr, of the same order as the rising time of the bubble over 20 kpc. The successive jet episodes generate their own pair of cavities. The bubble can entrain some low-entropy ICM gas from the center of the cluster to higher radius where the AGN heating is less efficient. It is therefore at this distance from the center (~20 kpc) that most of the gas can cool. The radiative cooling takes a few tens of Myr, and the gas reaches the phase below 10^4 K (e.g. the high-resolution simulations by Revaz et al. 2008). The cold gas forming molecules, and detected in CO emission is then expected at the border of cavities, and along the filaments, as observed. The gas is no longer in pressure equilibrium, and should be infalling at the free-fall velocity. In fact, the kinematics observed in the CO emission lines show a much lower velocity, implying that the gas is slowed down by the ram pressure of the hot gas and/or the magnetic field (Olivares et al. 2019).

The molecular gas is well correlated with the Hα filaments, showing that the filaments are indeed the results of the cooling process. (Salomé et al. 2006) The ionized gas is easier to detect, and the filaments in Ha are more extended than in CO emission (Olivares et al. 2019). The physical processes at the origin of the gas cooling have been widely debated. When the ICM is simplified as a spherically symmetric, homogeneous medium, with one phase gas, the comparison between the cooling time $t_{\rm cool}$ and the free-fall time $t_{\rm ff}$ (computed as $(2r/g)^{1/2}$, with g the gravitational acceleration, derived from the mass $M(R)$ of the cluster), should give the regions where the gas is actually cooling. The ratio $t_{\rm cool}/t_{\rm ff}$ is always larger than 10 even at the cluster center (cf Figure 3.12). Simulations have shown however that the hot ICM can become thermally unstable locally, when the ratio of the cooling to free-fall timescales becomes equal to 10–20 (McCourt et al. 2012; Gaspari et al. 2012; Li et al. 2015; Voit et al. 2015, 2017). Other propositions is that low-entropy gas is uplifted by the bubbles up to a point where $t_{\rm cool}$ becomes smaller than its infall time at this location, that can be much longer than the free-fall time (McNamara et al. 2016).

In general, cool core clusters are relaxed and non-cool core clusters are still out of equilibrium due to a recent cluster merger. The cooling can still occur with a certain amount of sloshing. The X-ray image of Perseus cluster for instance, is lopsided at large scale, and reveals a one-arm spiral structure (Fabian et al. 2006). These perturbations are frequent in clusters, either due to a small group merging with the cluster, or to the BCG oscillating at the bottom of the potential. They can create cold fronts (Zuhone & Roediger 2016). In the peculiar cluster Abell 1795, a very extended Hα filament suggests the existence of a cooling wake, produced by a symmetric cooling flow falling onto the brightest cluster galaxy, which has a high relative velocity at the center of the cluster (Crawford et al. 2005). The cooling wake is traced by a long (46 kpc) filament bright in X-ray, Hα, and also CO emission. The latter is curling around the cavities dug by the radio jet (Russell et al. 2017). The cooling

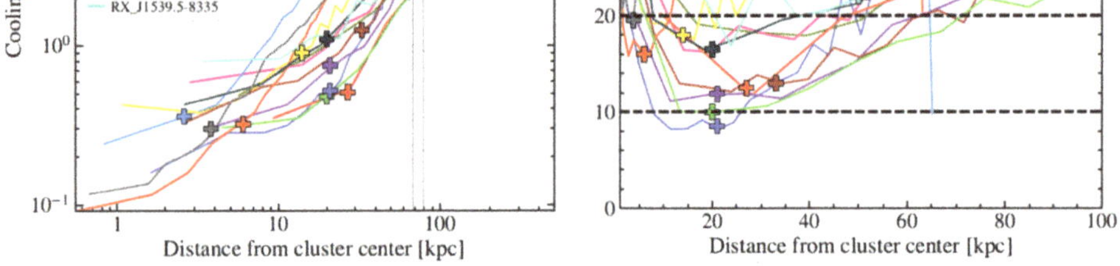

Figure 3.12. Left panel: cooling time as a function of distance from the center for several cool core clusters (labeled in the top left corner). The radial extent of the filaments is indicated with a cross and color-coded for each source. The value $t_{\rm cool}$ = 5 Gyr is highlighted by an horizontal dot-dashed line. Right panel: ratio of the cooling time to free-fall time, $t_{\rm cool}/t_{\rm ff}$ as a function of radius. The horizontal dashed lines are $t_{\rm cool}/t_{\rm ff}$ = 10 and 20, the approximate threshold for the onset of thermal instabilities (e.g. Li et al. 2015). Reproduced from Olivares et al. (2019). © ESO.

timescale is of the same order (300 Myr) as the oscillation period of the BCG in the cluster potential.

Several theoretical works have clarified the feedback process and the nature of filaments (e.g. Gaspari et al. 2017; Voit 2021). The X-ray atmosphere of galaxy clusters can be considered as a multiphase medium in constant evolution, with cold gas raining down onto the central BCG. The AGN heating through radio jets is stirring up the medium, creating subsonic turbulence, of dispersion ~150 km s^{-1}, as measured in the hot gas at 30–60 kpc from the center by the Hitomi Collaboration (2016). The turbulent eddies create overdensities, triggering the hot gas cooling even far from the center. This thermal instabilities occur when $t_{\rm cool}/t_{\rm ff}$ = 10–20, and produce a cascade of multiphase gas filaments, raining down onto the central galaxy. The filaments contain cold molecular gas clumps, with an ionized interface, producing Hα emission. This raining down of cold material fuels the central black hole, and is regulated by the AGN feedback. This process can be view as a precipitation of cold gas from the hot atmosphere (Voit 2021), maintaining the $t_{\rm cool}/t_{\rm ff}$ ratio constant throughout the core. When the cold gas clumps arrive in the central 1 kpc they collide with each other, in a more or less inelastic way, exchanging away angular momentum and producing chaotic cold accretion (Gaspari et al. 2017). These considerations are supported by observations of the kinematics in filaments (cf Figure 3.13). There is both inflow and outflow, due to the chaotic motions, but in the final stage of the raining down, the cold molecular gas inflow towards the BCG dominates, as seen in molecular gas redshifted absorption in front of the AGN core

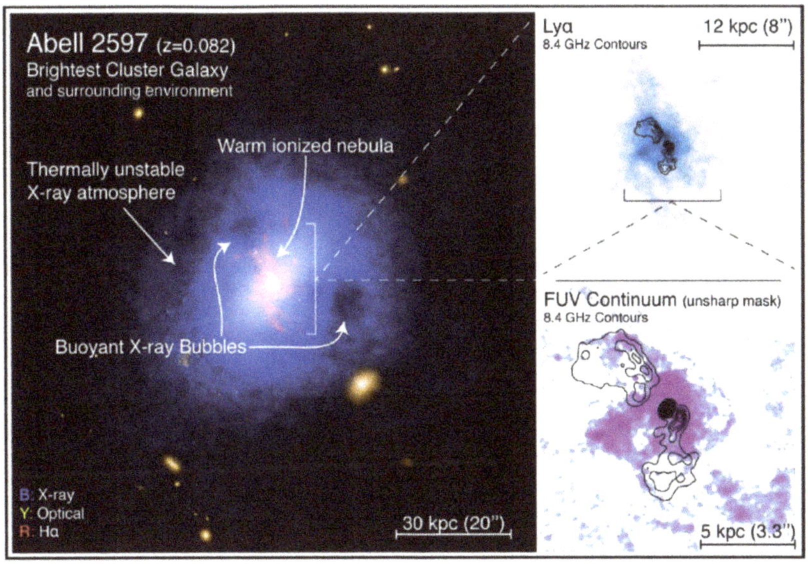

Figure 3.13. The Abell 2597 galaxy cluster and its BCG. Left: a superposition of X-ray (blue), optical (yellow) and Hα +[NII] (red) showing the bubbles and the filaments. Right: the Lyα (top) and FUV continuum (bottom) in color, with the radio contours superposed. Reprinted from Tremblay et al. (2018).

radio source (e. g. Tremblay et al. 2016, Rose et al. 2019). Globally, the jetted AGN feedback inflating bubbles, uplifting some warm gas up, which cools at the dense border of cavities, to rain down to the black hole, forms a galactic fountain (Tremblay et al. 2018). This process is similar to the fountain in galaxy-sized haloes, where gas ejected by supernovae, comes back to the galactic plane with more gas, after having triggered the cooling of the lower corona (Fraternali et al. 2013).

Observations of CO absorption in front of the AGN radio cores are illustrated in Figure 3.14. Similar phenomena are occurring in clusters and groups. The hot atmosphere of the group central galaxy NGC 5044 is also multiphase, and giant molecular associations are observed associated to H α filaments (David et al. 2014).

Finally, the order of magnitude of energy required from the AGN feedback to moderate the cooling flow is quite compatible with what is necessary from the observations. The kinetic power in the jets can be estimated from the product of the volume of the bubbles, and the surrounding pressure (obtained from the density and temperature of the thermal gas), divided by the buoyancy time (which depends on the gravitational potential). It can be seen that the radiative losses of the gas in the cooling zones are compensated by thermal energy plus the mechanical energy of cavity expansion (PV), cf Figure 3.15. The enthalpy is equal to 4 pV for a relativistic equation of state (with $\gamma = 4/3$). The cavities are emptied of their hot gas, the X emission comes only from the shell surrounding the cavities. After de-projection, the brightness of the cavities is compatible with the emission from the surface alone. The mass of the shells is comparable to the mass ejected from the cavities. The X gas was

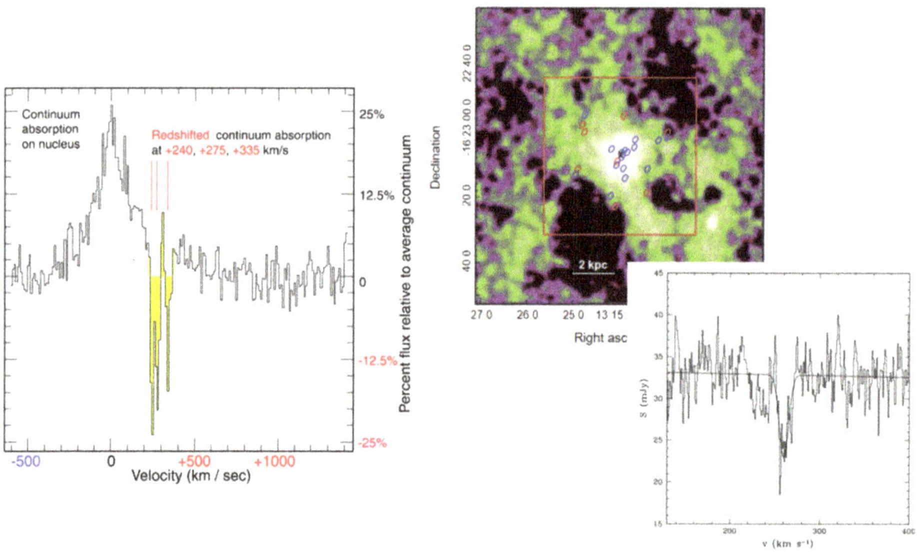

Figure 3.14. Left: the CO spectrum in front of the AGN core in Abell 2597 central galaxy: there are three components of redshifted absorption, implying inflow of gas. Reproduced from Tremblay et al. (2018). © 2018. The American Astronomical Society. All rights reserved. Right: the group central galaxy, NGC 5044, seen in X-ray, with the location of giant molecular complexes, and below the CO absorption spectrum towards the core. Reprinted from David et al. (2014). © 2014. The American Astronomical Society. All rights reserved.

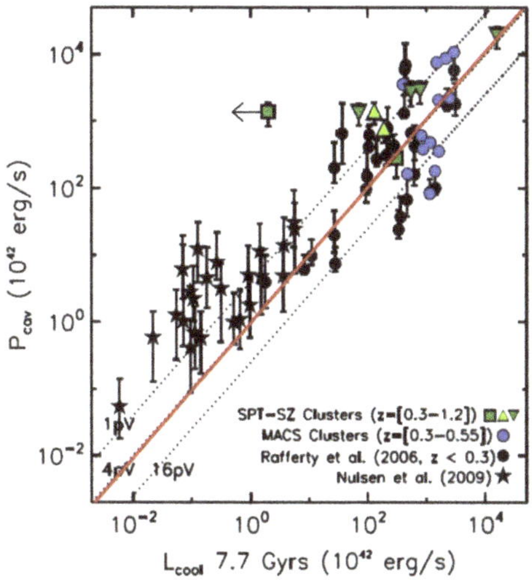

Figure 3.15. Cavity radio power P_{cav} plotted against X-ray luminosity radiated from the region where the cooling time is shorter than 7.7 Gyrs. The points correspond to several samples of clusters and groups. The red line corresponds to the energy required to make a cavity, i.e., thermal energy + PV work to inflate cavity, which is 4 PV for relativistic gas. Reprinted from Hlavacek-Larrondo et al. (2015).

pushed out of the cavities by the plasma ejected by the radio source and compressed into the shells (see Fabian 2012).

Most of the central galaxies in cool core clusters are radio-loud AGN, as was mentioned earlier. To probe the AGN feedback impact, it is important to correlate the radio emission with signatures of hot gas cooling. The amount of cooling gas maybe quantified by the optical emitting lines, Hα + [NII] in filaments, implying the existence of multiphase and cooling gas. However, there might be different time-scales for radio emission, and cooling of the gas at 30 kpc scale. It is then important to determine the age of the radio source, for instance through the slope of the synchrotron emission and the detailed radio spectrum. The brightest radio cores are hosted exclusively by the most X-ray luminous clusters, as established by Hogan et al. (2015), see Figure 3.16. It appears that BCG in line emitting clusters are much radio powerful than in non-line emitting ones, and their duty-cycle is higher. Figure 3.16 shows a correlation between the core radio power and the cavity power, which is only expected if the duty-cycle of the core is at least equal to the bubble inflating and rising timescale (300Myr).

3.5 How Efficient is the AGN Feedback?

Observations show that the AGN feedback is efficient in at least two circumstances: in galaxy clusters, where the AGN radio mode is able to moderate and regulate the cooling flows, as was just discussed, and in individual galaxies, at the high-luminosity end of the AGN (either in radiative or radio mode, cf Section 4.1). The outflowing gas is reaching escape velocity only in less than 5% of cases, and the

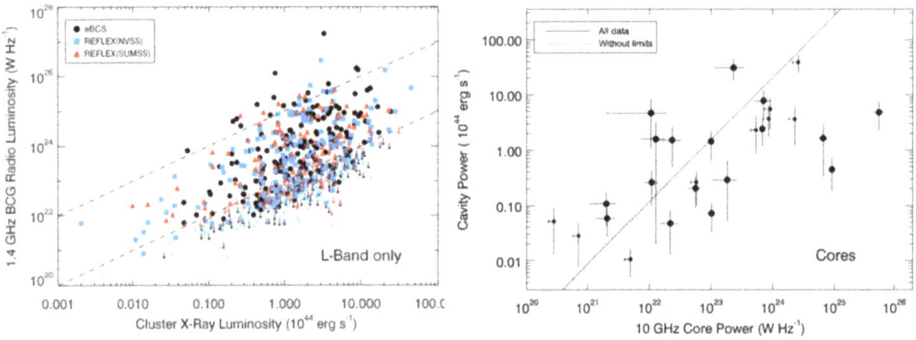

Figure 3.16. Left: radio luminosity of the BCG galaxy at 1.4 GHz, versus its cluster X-ray luminosity. Down arrows indicate upper limits of the radio flux. Consequently, the lack of points in the bottom right is due to sensitivity limitations. The dashed lines have a slope of 1, and separated by 2 orders of magnitude, showing the large scatter. Right: the X-ray cavity power as a function of core radio emission. The fit over all data is the solid line, while the dotted line is the linear fit when the upper limits on the radio-data are neglected. Reproduced from Hogan et al. (2015).

depletion timescale of the interstellar medium is only marginally affected, or only in the central region (Fluetsch et al. 2019).

Simulations have studied AGN feedback, with various methods, and with several hypotheses, and resolutions. The results present a wide range of conclusions, from negative to positive feedbacks. In their simulation of the AGN jet feedback on a galactic disk, Gaibler at al. (2012) resolved the jet and the clumpy multiphase interstellar medium, and used star formation sub-grid recipes. After 10 Myr, the jet has dug a cavity in the central region of the galaxy disk, but in the same time, created a bubble that shocked and compressed the gas in the disk, yielding a ring of star formation. Progressively the shock wave propagates in the galaxy disk, and the whole is involved by the triggered star formation, showing that AGN feedback can be positive.

Feedback can come from supernovae, AGN or both, and simulations have to tune sub-grid parameters, to take into account their effects on resolved scales, and calibrate them with observations. The results have been quite successful to reproduce the properties of galaxies, their stellar content, and the relation between the black hole and stellar mass of galaxies. It is important then to have further constraints, essentially the luminosity function across redshift, and the AGN luminosity function. According to resolution and processes considered, the relative importance of galaxy mergers, and cold gas accretion for black hole growth and star formation quenching varies. In the IllustrisTNG simulations, Weinberger et al. (2018) consider the BH growth of massive galaxies, and select to reinject thermal energy in the surrounding medium, for high Eddington ratios (radiative mode) and kinetic energy for low ratios (the threshold depending on the BH mass). At high redshift the supernovae feedback dominates, but rapidly the AGN will take over. The radiative mode is certainly more powerfull, however its efficiency is reduced by the high cooling rate of the dense gas around. The AGN kinetic mode is therefore dominant in the quenching of massive galaxies. The main growth of BH are then in the radiative mode, which is visible in the luminosity function. The main feedback and quenching efficiency is in the kinetic mode, where gas inflow is impeded, and the BH growth is regulated. Growth then occurs at $z = 0$ through dry mergers and BH coalescence (Weinberger et al. 2018). A similar conclusion was also obtained in SPH simulations by Choi et al. (2015), who reproduced the M–σ relation for black hole masses, and conclude that the mechanical feedback is the most efficient to quench star formation (Figure 3.17). With the AMR code RAMSES, Cielo et al. (2018) have tested the radiative and kinetic modes, allowing the jet plasma to expand in a clumpy fractal cold medium in galaxy disks. Radiative mode and jets perpendicular to the disk do not lead to efficient coupling with the interstellar medium, and have minimal feedback, while the kinetic mode with jets inclined towards the disk are the most efficient to drive massive outflows, and regulate star formation.

These conclusions might depend on the selected process for feedback, either thermal or kinetic energy. Powerful AGN are able to ionize the gas far away in the galactic disk. Roos et al. (2015) have studied in particular this process, through radiative transfer in post-processing of high spatial resolution simulations (cf Figure 3.18). However, molecular clouds in the disk are too dense to be ionized. The results show that the SFR is only reduced by a few percent at most by taking into account ionization, and therefore the long-range radiative effects are negligible, although the AGN was strong enough to drive powerful outflows.

High-resolution simulations with the AMR code RAMSES have also shown how the AGN feedback could be positive, and trigger star formation (Bieri et al. 2016). They investigate the effect of an ambient external pressure on molecular clouds, which could be due to AGN heating, and also cosmic rays, accelerated in shocks, produced by the AGN jet. Even moderate over-pressure leads to more

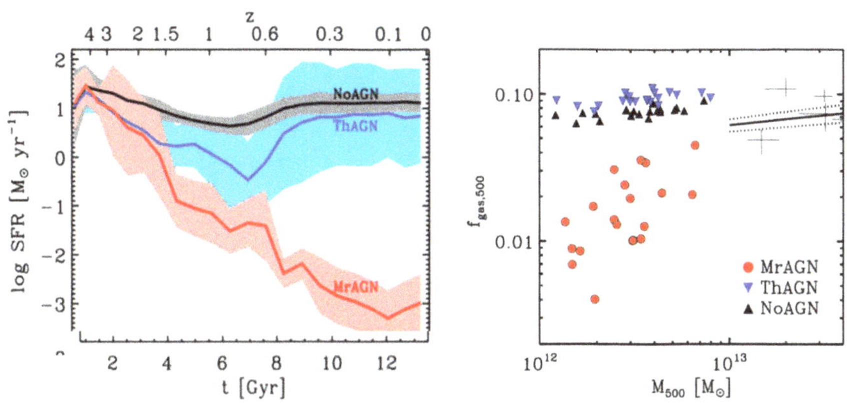

Figure 3.17. Simulations of the impact of AGN feedback on massive galaxies. Left: averaged star formation rate over time for the central galaxies for the model without AGN feedback (NoAGN, black), thermal feedback (ThAGN, blue), and momentum and radiative feedback (MrAGN, red) measured within 10% of the virial radii. The shading represents 1σ scatter. Right: gas mass fraction versus total mass, for the three feedback models. Gas mass fraction and total mass are measured within r500, the radius where the spherical over-density in the galaxy drops below 500 times the critical density of the universe at $z = 0$. Reproduced from Choi et al. (2015).

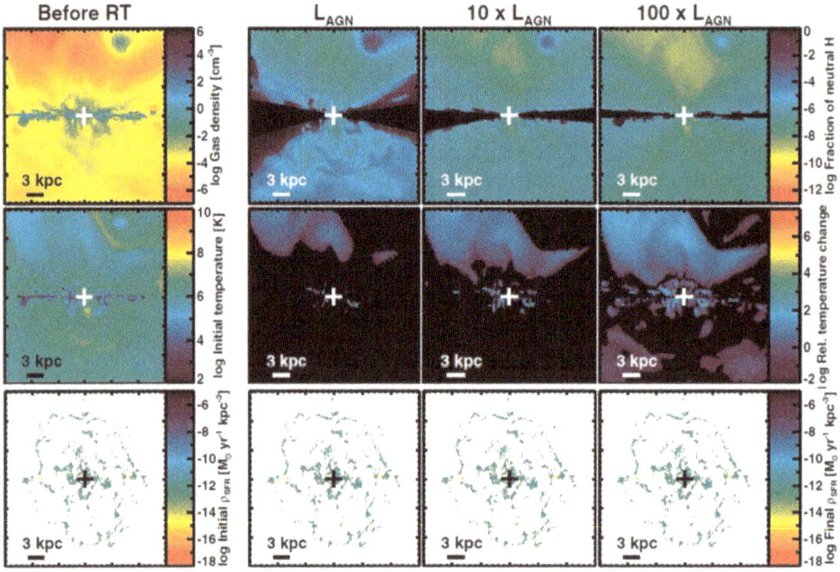

Figure 3.18. Simulations of the AGN feedback on a disk galaxy, both thermal feedback, and due to ionizing radiation, through a high-resolution radiative transfer code, by Roos et al. (2015). Left column before radiative transfer (RT), then for three values for the AGN luminosity. The first and second rows are the density and temperature, for the edge-on disk (left), and then neutral gas fraction and change in temperature. The third row is the star formation rate for the face-on disk. Reprinted from Roos et al. (2015).

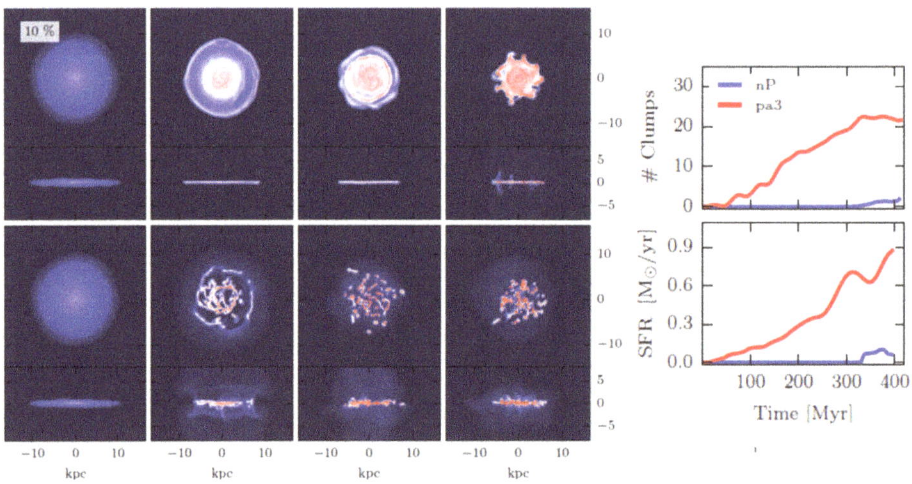

Figure 3.19. Left: face-on and edge-on maps of the gas density of two simulations without enhancement of the external pressure (model nP, top row), and with enhancement (model pa3, bottom row). The four columns correspond to the different snapshot at $T = 0$, 50, 200 and 400 Myr. Right: number of clumps (top) and star formation rate (bottom) as a function of time, for the nP model (blue) and pa3 model (red). Increasing pressure outside the galaxy leads to enhanced clump and star formation. Reproduced from Bieri et al. (2016).

fragmentation of the gas, and triggers star formation (cf Figure 3.19). A boost of star formation by an order of magnitude may be the consequence of the AGN feedback.

Bibliography

Aalto, S., Costagliola, F., Muller, S., et al. 2016, A&A, 590, A73
Aalto, S., Falstad, N., Muller, S., et al. 2020, A&A, 640, A104
Baugh, C. M. 2006, RPPh, 69, 3101
Behroozi, P. S., Wechsler, R. H., & Conroy, C. 2013, ApJ, 70, 57
Best, P. N., von der Linden, A., Kauffmann, G., et al. 2007, MNRAS, 379, 894
Bieri, R., Dubois, Y., Silk, J., et al. 2016, MNRAS, 455, 4166
Canning, R. E. A., Ryon, J. E., Gallagher, J. S., et al. 2014, MNRAS, 444, 336
Cecil, G., Greenhill, L. J., DePree, C. G., et al. 2000, ApJ, 536, 675
Choi, E., Ostriker, J. P., Naab, T., et al. 2015, MNRAS, 449, 4105
Cicone, C., Maiolino, R., Sturm, E., et al. 2014, A&A, 562, A21
Cielo, S., Bieri, R., Volonteri, M., et al. 2018, MNRAS, 477, 1336
Conselice, C. J., Gallagher, J. S. III, & Wyse, R. F. G. 2001, AJ, 122, 2281
Costa, T., Sijacki, D., & Haehnelt, M. G. 2014, MNRAS, 444, 2355
Crawford, C. S., Sanders, J. S., & Fabian, A. C. 2005, MNRAS, 361, 17
Crenshaw, D. M., & Kraemer, S. B. 2000, ApJ, 532, L101
Crenshaw, D. M., Kraemer, S. B., & George, I. M. 2003, ARA&A, 41, 117
Dasyra, K., & Combes, F. 2012, A&A, 541, L7
Dasyra, K., Combes, F., Oosterloo, T., et al. 2016, A&A, 595, L7
David, L. P., Lim, J., Forman, W., et al. 2014, ApJ, 792, 94
Edge, A. 2001, MNRAS, 328, 762
Eke, V. R., Baugh, C. M., Cole, S., et al. 2006, MNRAS, 370, 1147
Fabian, A. C., Sanders, J. S., Allen, S. W., et al. 2003, MNRAS, 344, L43
Fabian, A. C., Sanders, J. S., Taylor, G. B., et al. 2006, MNRAS, 366, 417
Fabian, A. C. 2012, ARA&A, 50, 455
Faucher-Giguère, C.-A., & Quataert, E. 2012, MNRAS, 425, 605
Ferland, G. J., Fabian, A. C., Hatch, N. A., et al. 2009, MNRAS, 392, 1475
Feruglio, C., Maiolino, R., Piconcelli, E., et al. 2010, A&A, 518, L155
Fiore, F., Feruglio, C., Shankar, F., et al. 2017, A&A, 601, A143
Fluetsch, A., Maiolino, R., Carniani, S., et al. 2019, MNRAS, 483, 4586
Fraternali, F., Marasco, A., Marinacci, F., & Binney, J. 2013, ApJ, 764, L21
Gabor, J. M., & Bournaud, F. 2014, MNRAS, 441, 1615
Gaibler, V., Khochfar, S., Krause, M., & Silk, J. 2012, MNRAS, 425, 438
Gallimore, J. F., Henkel, C., Baum, S. A., et al. 2001, ApJ, 556, 694
Gallimore, J. F., Baum, S. A., & O'Dea, C. P. 2004, ApJ, 613, 794
García-Burillo, S., Combes, F., Usero, A., et al. 2014, A&A, 567, A125
García-Burillo, S., Combes, F., Ramos Almeida, C., et al. 2016, ApJ, 823, L12
García-Burillo, S., Combes, F., Ramos Almeida, C., et al. 2019, A&A, 632, A61
Gaspari, M., Ruszkowski, M., & Sharma, P. 2012, ApJ, 746, 94
Gaspari, M., Temi, P., & Brighenti, F. 2017, MNRAS, 466, 677
Ghisellini, G., Della Ceca, R., Volonteri, M., et al. 2010, MNRAS, 405, 387
Gnedin, N. Y. 2000, ApJ, 542, 535
Gratadour, D., Rouan, D., Grosset, L., et al. 2015, A&A, 581, L8

Gravity Collaboration, 2020, A&A, 634, A1
Helou, G., Soifer, B. T., & Rowan-Robinson, M. 1985, ApJ, 298, L7
Hitomi Collaboration, 2016, Nature, 535, 117
Hlavacek-Larrondo, J., McDonald, M., Benson, B. A., et al. 2015, ApJ, 805, 35
Hogan, M. T., Edge, A. C., Hlavacek-Larrondo, J., et al. 2015, MNRAS, 453, 1201
Hopkins, P. F., & Elvis, M. 2010, MNRAS, 401, 7
Hughes, S. A., & Blandford, R. D. 2003, ApJ, 585, L101
Imanishi, M., Nguyen,, Dieu, D., Wada, K., et al. 2020, ApJ, 902, 99
Impellizzeri, C. M. V., Gallimore, J. F., Baum, S. A., et al. 2019, ApJ, 884, L28
King, A. 2003, ApJL, 596, L27
King, A. R., Pringle, J. E., & Hofmann, J. A. 2008, MNRAS, 385, 1621
King, A. R., Zubovas, K., & Power, C. 2011, MNRAS, 415, L6
King, A., & Pounds, K. 2015, ARA&A, 53, 115
King, A. 2016, MNRAS, 456, L109
Li, Y., Bryan, G. L., Ruszkowski, M., et al. 2015, ApJ, 811, 73
Lim, J., Ohyama, Y., Chi-Hung, Y., et al. 2012, ApJ, 744, 112
McConnell, N. J., Ma, C.-P., Gebhardt, K., et al. 2011, Natur, 480, 215
McCourt, M., Sharma, P., Quataert, E., & Parrish, I. J. 2012, MNRAS, 419, 3319
McDonald, M., Bayliss, M., Benson, B. A., et al. 2012, Natur, 488, 349
McNamara, B. R., Russell, H. R., Nulsen, P. E. J., et al. 2016, ApJ, 830, 79
Morganti, R., Frieswijk, W., Oonk, R J B., et al. 2013a, A&A, 552, L4
Morganti, R., Fogasy, J., Paragi, Z., et al. 2013b, Sci, 341, 1082
Morganti, R., & Oosterloo, T. 2018, A&ARv, 26, 4
Mukherjee, D., Wagner, A. Y., Bicknell, G. V., et al. 2018, MNRAS, 476, 80
Murray, N., Quataert, E., & Thompson, T. A. 2005, ApJ, 618, 569
Olivares, V., Salomé, P., Combes, F., et al. 2019, A&A, 631, A22
Peterson, J. R., & Fabian, A. C. 2006, PhR, 427, 1
Reeves, J. N., & Braito, V. 2019, ApJ, 884, 80
Revaz, Y., Combes, F., & Salomé, P. 2008, A&A, 477, L33
Richings, A. J., & Faucher-Giguère, C.-A. 2018a, MNRAS, 474, 3673
Richings, A. J., & Faucher-Giguère, C.-A. 2018b, MNRAS, 478, 3100
Rodríguez-Ardila, A., Prieto, M. A., Viegas, S., & Gruenwald, R. 2006, ApJ, 653, 1098
Roos, O., Juneau, S., Bournaud, F., & Gabor, J. M. 2015, ApJ, 800, 19
Rose, T., Edge, A. C., Combes, F., et al. 2019, MNRAS, 489, 349
Russell, H. R., McNamara, B. R., Fabian, A. C., et al. 2017, MNRAS, 472, 4024
Russell, H. R., McNamara, B. R., Fabian, A. C., et al. 2019, MNRAS, 490, 3025
Sądowski, A., Narayan, R., Penna, R., & Zhu, Y. 2013, MNRAS, 436, 3856
Salomé, P., & Combes, F. 2003, A&A, 412, 657
Salomé, P., Combes, F., Edge, A. C., et al. 2006, A&A, 454, 437
Schinnerer, E., Eckart, A., Tacconi, L. J., et al. 2000, ApJ, 533, 850
Schmitt, H. R., & Kinney, A. I. 2002, NewAR, 46, 231
Silk, J., & Rees, M. J. 1998, A&A, 331, L1
Spoon, H. W. W., Farrah, D., Lebouteiller, V., et al. 2013, ApJ, 775, 127
Tombesi, F., Cappi, M., Reeves, J. N., et al. 2010, A&A, 521, A57
Tremblay, G. R., Oonk, J. B. R., Combes, F., et al. 2016, Nature, 534, 218
Tremblay, G. R., Combes, F., Oonk, J. B. R., et al. 2018, ApJ, 865, 13

Veilleux, S., Meléndez, M., Sturm, E., et al. 2013, ApJ, 776, 27
Veilleux, S., Meléndez, M., Tripp, T. M., et al. 2016, ApJ, 825, 42
Veilleux, S., Maiolino, R., Bolatto, A. D., & Aalto, S. 2020, A&ARv, 28, 2
Vermot, P., Clénet, Y., & Gratadour, D. 2019, A&A, 629, A98
Voit, G. M., Donahue, M., Bryan, G. L., & McDonald, M. 2015, Natur, 519, 203
Voit, G. M., Meece, G., Li, Y., et al. 2017, ApJ, 845, 80
Voit, G. M. 2021, ApJ, 908, L16
Wagner, A. Y., Bicknell, G. V., Umemura, M., et al. 2016, AN, 337, 167
Walker, S. A., Fabian, A. C., Russell, H. R., & Sanders, J. S. 2014, MNRAS, 442, 2809
Weinberger, R., Springel, V., Pakmor, R., et al. 2018, MNRAS, 479, 4056
Weymann, R. J., Morris, S. L., Foltz, C. B., & Hewett, P. C. 1991, ApJ, 373, 23
White, S. D. M., & Rees, M. J. 1978, MNRAS, 183, 341
White, S. D. M., & Frenk, C. S. 1991, ApJ, 379, 52
Zubovas, K., & King, A. R. 2014, MNRAS, 439, 400
Zuhone, J. A., & Roediger, E. 2016, JPlPh, 82, 535820301

Chapter 4

The Circumnuclear Region

The dusty torus with a doughnut shape introduced by the active galactic nuclei (AGN) unification model is progressively replaced by dusty outflows, producing the required obscuration. This hot dusty component is radiating in the near and mid-infrared, and is producing a bump in the spectral energy distribution (SED) of the global AGN host, with the characteristic silicate signature. Their outflowing component has been successfully modeled by polar bicones or hourglass-shaped structures. ALMA has discovered in nearby Seyferts molecular tori, a relatively thin disk, decoupled by morphology and kinematic from the rest of the galaxy. The molecular tori have typical radii 10–20 pc, and masses a few $10^7\ M_\odot$. The origin of their misalignment is discussed.

4.1 The Necessity of a Dusty Torus

According to the widely accepted paradigm for active galactic nuclei (AGN), geometry and orientation with respect to the line of sight are two fundamental parameters of the unification model (Antonucci & Miller, 1985), already described in Section 1.8 of Chapter 1. The two main classes, AGN of type 1 and 2 (with or without broad line region or BLR) can be explained if the central BLR is hidden or obscured in the type 2 objects. The obscuration must be provided by a dusty torus, a thick and roughly axisymmetric structure of 1–10 pc size typically, just outside the dust-free accretion disk and BLR. For luminous AGN (at least 0.01 Eddington luminosity), there exists a thin optically thick rotating accretion disk, feeding efficiently the central black hole (BH), below 1 pc size. At larger scales (10–100 pc and beyond), outside the torus, multiphase gas is observed in optical or millimetric lines, with lower velocities (narrow line region, or NLR).

The required torus should be quite thick, with height comparable to radius, if we believe the statistical relative number of Seyfert 1 and 2 (e.g., Elitzur & Shlosman 2006). This led to the popular doughnut morphology. The center of the torus is depleted due to the evaporation of the gas by the active nucleus, and the launching of

doi:10.1088/2514-3433/ac2a27ch4

a wind. The torus must have column densities high enough to completely obscure optically the BLR region. This means at least a gas column density of 10^{23} cm^{-2}, depending on the gas metallicity, and dust-to-gas ratio. At these columns, some of the X-ray radiation can still be seen across, however, at 10^{24} cm^{-2}, even the 2–10 keV X-rays are obscured, and at 10^{25} cm^{-2}, the X-rays above 10 keV also. These AGN are called mildly and heavily Compton-thick, respectively. The torus must be thick, and this requires a high-velocity dispersion (50–100 km s^{-1}) especially in the vertical direction, to maintain this thickness. The gas cannot be homogeneous, since then these velocities will be equivalent to a temperature much too high (~10^6 K) and the dust would not survive. The torus must then be clumpy, fragmented in dense clouds, with a covering factor of order unity, and a total column density in the equatorial plane of 10^{24} cm^{-2}, through superposition of clouds along the line of sight (Krolik & Begelman 1988). To maintain the thickness of the torus, the cloud collision rate must be relatively low. Clouds are heated and evaporate their surface on the side facing the nucleus, but the dense cold gas in the center of the clouds is shielded from the AGN. Clouds must be dense to survive the strong tidal shearing of the Keplerian potential; they are marginally self-gravitating, are partially sheared and re-form through collisions, and are compressed by the high pressure imposed by the central AGN. The collisions, although at a low rate, provide the dissipation necessary for the clouds to spiral in toward the center and feed the nucleus. The issue has long been debated to determine the heating source able to maintain the high vertical velocity dispersion, and the high thickness. Star formation processes are not energetic enough, and the only solution could be that the torus is not stable but represents intermittent accretion episodes (Collin & Zahn 2008). Numerical simulations succeed to form relatively thick disks in the outer parts, due to the supernova feedback, radiation pressure and winds from the AGN (Wada et al. 2009). N-body simulations of the numerous ($N \sim 10^5$) clumps in the torus show that the torus keeps its thickness, and the highest velocity dispersions come from the torus throat (Bannikova et al. 2021).

High-resolution interferometric observations in the mid-infrared with the Very Large Telescope Interferometer (VLTI-MIDI) and the Keck of nearby Seyferts have shown that the dust at 10 pc-scale is not mainly distributed in a torus structure in the plane, but on the contrary, is elongated in the perpendicular direction (e.g., Tristram et al. 2014; Asmus et al. 2016; Asmus 2019). This partly solves the issue of the thickness of the torus: the dust is heated by the nucleus starting at the sublimation radius, and is entrained in a wind in the perpendicular direction by radiation pressure on dust. This forms a hollow polar cone, as described schematically in Figure 4.1.

The sublimation radius is the location where dust grains evaporate due to high temperatures. This temperature is 1400K for silicate grains and 1800K for graphite ones. The sublimation radius varies with the AGN luminosity as $L^{1/2}$, and is of the order of 1 pc for a typical nearby Seyfert. The disk continues at larger radii (10–100 pc) in the equatorial plane, in a molecular torus, which is much thinner, confined to the plane. At these distances, the disk is unstable to fragmentation, and may form stars (according to the Toomre-Q parameter, e.g., Collin & Zahn 2008). The dust

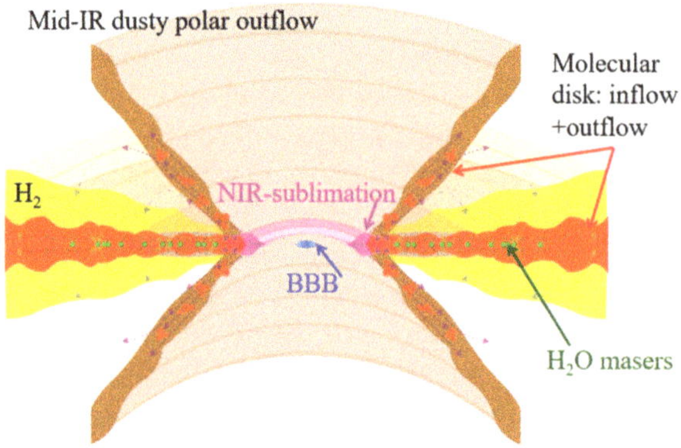

Figure 4.1. Schematic view of the molecular and dusty circumnuclear region of an active nucleus. The central big blue bump region (BBB) is the Keplerian broad line region (BLR). In pink is delineated the ring at dust sublimation radius, source of most near-IR emission. The brown features are the molecular and dusty disk plus the biconical polar outflows, at the walls of the hollow polar cone. This is the region of most mid-IR emission, all millimeter molecular lines, water masers and the H_2 near-IR lines. Arrows indicate the main stream flows of the gas. Reprinted from Hönig (2019). © 2019. The American Astronomical Society. All rights reserved.

and gas entrained in the wind are aligned with the ionized gas seen in [O III], [O IV], and is part of the narrow line region. At some altitude above the plane, depending on their velocity reached, the dust and gas may rain down on the disk, to continue to feed the AGN (e.g., Hönig 2019). The bulk of the mid-IR emission of the dust is aligned with the polar cones, which are ubiquitous in nearby AGN (cf Figure 4.2)

4.2 Dusty Torus SED

In order to confront all debated dusty torus models to observations, it is necessary to predict the spectral energy distribution (SED) expected from them. The predictions of radiative transfer models are somewhat easy within the hypothesis of an homogeneous medium, and they have been done analytically at 1 dimension in a clumpy medium (Nenkova et al. 2008a, 2008b). The main difference from a homogeneous radial density is that in a clumpy medium, a large range of dust temperatures coexist at the same distance from the AGN. In a homogeneous torus, the temperature decreases as the flux varies as r^{-2} and there is a very deep silicate absorption at 9.7 μm, which is not observed, in NGC 1068 for example. The more realistic 3D clumpy torus is more difficult to deal with, and has to be done entirely numerically (e.g., Stalevski et al. 2012). Many parameters can be varied, and all have an impact on the near- and mid-infrared SED: the clump size, their optical depth, the distribution in the torus. The silicate feature may be seen in emission in low optical depth tori, while in absorption with high-density clumps in the inner regions, and the shape of the profile depends strongly on the inclination. Two phase models, with dense clumps but also a more diffuse interclump medium, bring more freedom to reproduce the observations: The interclump medium has a high weight in the silicate feature emission and near-IR emission in general (cf Figure 4.3). Finally, for less nearby galaxies, the torus emission is diluted in the global spectrum of the galaxy

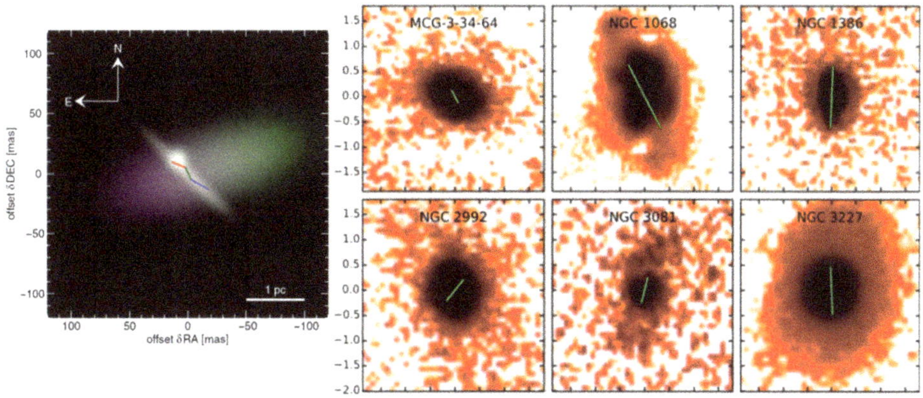

Figure 4.2. Left: mid-IR image of the nucleus of the Circinus galaxy, obtained with the interferometer VLTI-MIDI. The emission model contains three components: the logarithmic-scaled colors red, green and blue correspond to 13.0, 10.5, and 8.0 μm, respectively. There is a clear color gradient of the polar extended component due to the increase in the silicate depth toward SE. This excess of absorption shifts the luminosity center to the NW. The polar component is more diffuse, but its extension compensates, and it represents 80% of the emission. The blue–red line in the middle represents the maser disk (blue and redshifted sides). Reproduced with permission from Tristram et al. (2014). © ESO. Right: some mid-IR images of the nucleus of nearby Seyferts. The green lines indicate the polar axis of the accretion disks, or the jet directions. Reprinted from Asmus et al. (2016). © 2016. The American Astronomical Society. All rights reserved..

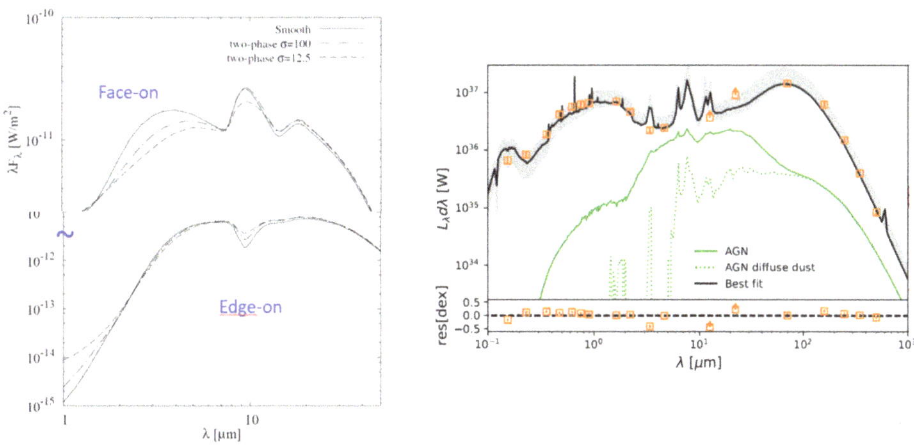

Figure 4.3. Left: simulations in 3D of the spectral energy distribution (SED) for a clumpy dusty torus. A smooth distribution (full line) is compared with two-phase clumpy media. σ is the ratio of the torus radius to the clump size, $\sigma = 12.5$ (big clumps), $\sigma = 100$ (small clumps). Face-on (top) and edge-on (bottom) views are not at the same scale. Reproduced from Stalevski et al. (2012). © 2012. The Royal Astronomical Society. All rights reserved. Right: the fit of NGC 1068 central SED, when the AGN contribution is diluted into the galactic disk stars and interstellar matter, producing PAH features (mainly at 3.3, 6.2, 7.7 and 11.3μm). Data points are in orange, with triangles indicating lower limits. The AGN torus contribution is the green line, while the green dotted line corresponds to the AGN light reprocessed by the diffuse dust in the galaxy. Reproduced with permission from Viaene et al. (2020). © ESO.

(Figure 4.3 right; Viaene et al. 2020). The AGN introduces a characteristic bump in the NIR of the galaxy SED.

Applying these models to nearby galaxies, with high resolution to isolate the torus, it was possible to derive the number of clumps, their density and radial profile (e.g., Ramos-Almeida et al. 2014). Another way to constrain the torus extent, geometry and composition is to use reverberation mapping. Although black hole growth requires an AGN lifetime of 10^7–10^9 yr, AGN light is variable in time, with different timescales, including flickering at 10^5 yr (Schawinski et al. 2015), down to month and day timescales (Peterson, 2001). Monitoring the K-band (2.2 μm) flux of AGN will give information on the hotter, inner boundary of the torus. Reverberation mapping experiments in the NIR have been conducted now on two dozens of nearby AGN, and confirm that the inner torus region (the sublimation radius) is of the order of 1 pc scale, and varying with AGN luminosity as $L^{1/2}$ (e.g., Netzer 2015).

The X-ray absorption of AGN luminosity might not be uniquely due to the torus. Indeed, there are sometimes variability timescales of the order of days, short enough that the corresponding radial scales correspond to the BLR. A large column density of gas without dust is sufficient to obscure in X-rays. Eclipses of the AGN in X-rays have been observed in NGC 1365 by Risaliti et al. (2007). They have been explained by short passages of dense clouds in the BLR, with a comet-shape morphology. The clouds are shocked in their >1000 km s^{-1} motion, and lose their gas in a long tail, of size of the order of 1 AU behind the cloud head (Maiolino et al. 2010). This implies that clouds in the BLR must be quickly replenished.

Modeling X-ray spectra obtained with XMM-Newton on nearby Seyferts, it is possible to resolve spectral features in the BLR 6.7 keV iron Kα line, which indicates a width of ~0.1 c and show evidence of gravitational redshift. Nandra et al. (2007) identify in these broad lines an ubiquitous narrow feature or "core," which they interpret as coming from the torus. Indeed, the narrow velocity must come from a larger distance from the center, and an optically thick material. This could be neutral Compton reflection, followed by fluorescence in the torus, with a velocity width of a few 100 km s^{-1}.

4.3 Molecular Tori

While the high interferometric resolution, obtained in the mid-infrared with the VLTI, has revealed the dusty torus scale, now with the high sensitivity and the high interferometric resolution of ALMA in the mm and sub-mm domain, it is possible to detect and map molecular tori, and structures at 10 pc scale, inside the black hole sphere of influence of nearby AGN. Several projects have selected nearby Seyferts to observe their nucleus with a resolution down to 0.05″ (typically 5 pc in the nearby universe observed). This is possible especially while targeting high rotation levels of the CO molecule, typically CO(3–2) up to CO(6–5). The wide ALMA bandwidth allows to detect simultaneously dense gas tracers, such as HCO+ and HCN, perfectly fitted to dense nuclei. For particularly dense and compact nuclei, a very useful tracer is the vibrationally excited HCN (e.g., Aalto et al. 2015).

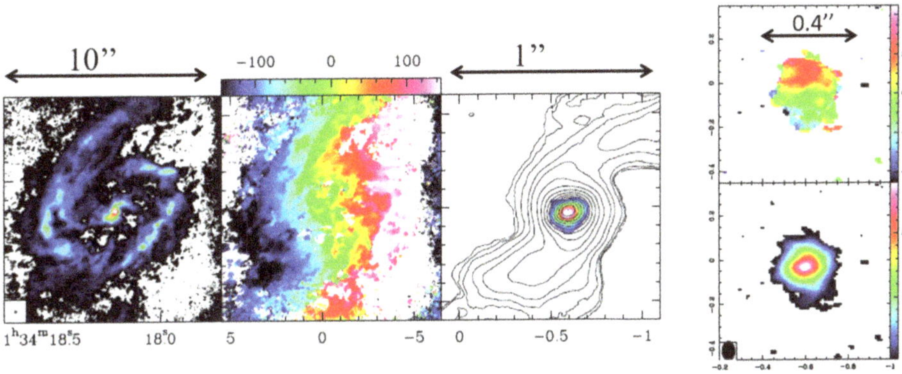

Figure 4.4. CO(3–2) images of NGC 613: integrated emission (moment zero), middle velocity field (moment one), and right superposition of the CO(3–2) contours to the continuum image (synchrotron from the AGN), with a zoom factor 10 with respect to the others. Right column, HCO+(4–3) velocity field (top) and integrated intensity (bottom), showing a decoupled disk of comparable size as the continuum emission. Reproduced with permission from Combes et al. (2019). © ESO.

An example of molecular torus is observed in NGC 613 (Figure 4.4). This galaxy, distant by 17 Mpc ($1'' = 83$ pc) is a late-type strongly barred spiral, with a star forming ring at the inner Lindblad resonance of the bar, at 350 pc in radius. The ring is encircling a secondary bar. Inside the ring, there is a relative depletion of gas, since gravity torques from the bar drive the gas towards the ring on both sides (Audibert et al. 2019). In the center of the ring, develops a trailing nuclear spiral, of radius $0.7'' \sim 60$ pc, which is responsible for the AGN feeding (see Chapter 2).

The nuclear spiral structure observed is the last before the molecular torus. Just inside this spiral, is embedded a compact circumnuclear disk, of radius 14 pc, which appears to be decoupled by its morphology and kinematics. Figure 4.4 shows that the kinematical major axis of the large-scale disk is −60°, while the torus has a kinematic major axis of PA=0 (in the HCO^+ velocity map) Their inclinations are also different (36° and 46° respectively, Combes et al. 2019). The torus is exactly centered on the continuum detected at 0.85 μm wavelength, which is essentially synchrotron emission from the AGN. This decoupled molecular disk has a mass of $3.9 \times 10^7\ M_\odot$, and an average surface density of $6 \times 10^4\ M_\odot/\mathrm{pc}^2$, or 7×10^{24} at/cm^2, which is enough to obscure an active nucleus, depending on the orientation of the line of sight, to a Compton-thick level. In general, these molecular disks are rather thin geometrically. However, they participate in their inner part to the molecular/dusty wind, and can contribute to the obscuration required by the unification model. In the particular case of NGC 613, a molecular outflow has been detected, in the same direction as the radio jet, perpendicular to the molecular torus (Audibert et al. 2019).

The size of the molecular torus depends on the molecular tracer used, suggesting that its density varies strongly with radius. For NGC 1068, one of the best studied Seyfert 2, the torus identified in CO(6–5) has a radius of 3.5 pc (Garcia-Burillo et al. 2016), while it is 14 pc in CO(2–1) (Garcia-Burillo et al. 2019). Water masers delineate the inner part of the torus, at 1 pc radius (Gallimore et al. 2004). The torus is located inside a circumnuclear ring of molecular gas, of radius 200 pc, which corresponds to the inner Lindblad resonance (ILR) of the nuclear bar. The torus and the AGN at the center are both displaced from the center of the ILR ring, by 70 pc. The origin of this lopsided morphology is not well known, but could be a consequence of the radio jet interaction with the gas in the disk. Here again, the molecular torus is decoupled from the rest of the disk, with a different position angle and inclination. The torus is almost edge-on, while the galaxy disk (including the ILR ring) has an inclination of 41°. The radio jet starts perpendicular to the accretion disk and torus, and then is deflected due to its interaction with the disk interstellar medium (ISM). The jet is much more extended in the NE direction, since it is propagating freely, after its deflection in a NE radio knot. While in the south, the jet is stopped by this interaction, and by forming a cocoon with shocks. In the south direction, the disk is particularly empty in neutral gas, over a 130 pc radius, may be pushed and/or ionized by the jet shocks. Two feedback mechanisms are combined in this galaxy, as more generally in many low-luminosity AGN, and a disk wind from the accretion disk is detected in ionized gas emission lines ([O III], Brγ, [Fe II]), as an ~1000 km s^{-1} outflow (Barbosa et al. 2014). This outflow is however symmetric with respect to the central engine, and only the radio jet could produce the asymmetry.

The molecular torus in NGC 1068 shows a very disturbed kinematics, averaging out in a surprisingly low rotation (Garcia-Burillo et al. 2014). This has been interpreted as counter-rotation in the outer parts of the torus, while the inner part (maser disk) is rotating in the same sense as the ILR ring and large-scale disk (Imanishi et al. 2020). This could also be the signature of the Papaloizou-Pringle instability (PPI) expected to occur in the nearly Keplerian potential of AGN tori (Papaloizou & Pringle 1984; PPI). Warps and slight misalignment are expected as a dynamical evolution, and the jet is indeed observed a bit tilted with respect to the molecular torus. This phenomenon is general, radio jets orientation are not random with respect to torus orientation, but are frequently tilted by less than 30° (Kamali et al. 2019). The PPI instability has been simulated for an isolated torus, and for tori perturbed due to the accretion of gas (Donmez 2014, 2017). These simulations show the growth of a long-lasting non-axisymmetric (lopsided) $m = 1$ mode, with vertical extensions. This instability may explain the lopsided distribution and noncircular motions in the gas, and might develop waves up to 10 pc scale.

Molecular tori have also been detected in early-type spirals, as is the case in the LINER NGC 1326, a strongly barred lenticular galaxy in the Fornax Cluster. It has an embedded nuclear bar, and a prominent ring of star formation, at the ILR of the primary bar (average radius of 300 pc). This ring dominates the star formation rate of the galaxy, so that 83% of the total Hα luminosity of NGC 1326 comes from the nuclear ring (Crocker et al. 1996). The molecular torus of radius 21 pc is almost edge-on, decoupled kinematically and morphologically from the ILR ring (cf Figure 4.5), and has a column density able to obscure an AGN to the Compton-thick level (Combes et al. 2019). The warm H_2 gas, detected at 2 μm by SINFONI at

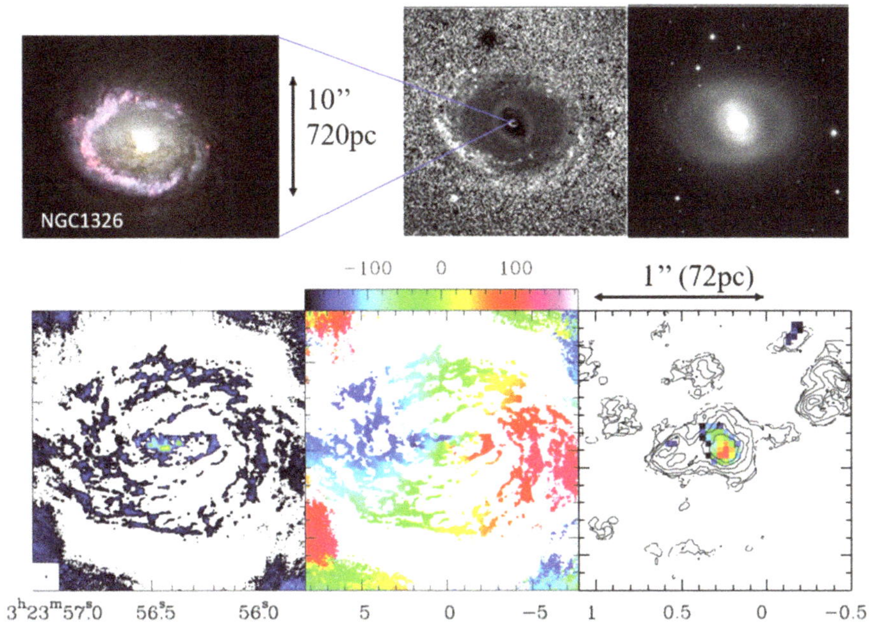

Figure 4.5. The molecular torus in the strongly barred lenticular galaxy NGC 1326. Top: the HST image of the star forming ring of external radius 5.7″= 410 pc, is zoomed out of the optical picture of the galaxy, with its unsharp masked image (Buta et al. 2000, and Carnegie-Irvine Galaxy Survey). Bottom: CO(3–2) integrated emission, velocity field, and at right superposition of the CO(3–2) contours to the continuum image (synchrotron from the AGN), with a zoom factor of 10. Reproduced with permission from Combes et al. (2019). © ESO.

the VLT, has a strong concentration in the torus, and should be excited essentially by shocks (Fazeli et al. 2020).

Another valuable example is provided by the Seyfert 2 galaxy NGC 5643, a barred late-type spiral, studied with ALMA by Alonso-Herrero et al. (2018). The circumnuclear region reveals a nuclear spiral, and inside this spiral a molecular torus of radius $r = 13$ pc, edge-on, while the main galaxy disk is almost face-on. The torus appears to collimate the high-velocity outflow of the NLR with Hα and [O III] emission lines. Its mass is $10^7\ M_\odot$, and it is Compton-thick. The ionization cone has its axis almost in the plane of the sky. The NLR is perpendicular to the torus and interact with the disk plane. The jet and NLR are co-spatial, the jet is extended about 2 kpc on each side of the torus. The jet–ISM interaction could produce shocks to accelerate the outflow.

All the molecular tori detected until now allow to derive some statistical trends (Combes et al. 2019; Garcia-Burillo et al. 2021). Their radii range from 10 to 20 pc, and they are inside the sphere of influence (SoI) of their central black hole. The radius of this sphere corresponds to the limit when the gravity of the bulge equals that of the BH, i.e., $R_{\rm SoI} = GM_{\rm BH}/\sigma^2$ of the order of ~20 pc for $M_{\rm BH} = 2 \times 10^8\ M_\odot$, and the central stellar velocity dispersion is $\sigma = 200$ km s^{-1}. When their kinematics is not perturbed by counter-rotating gas or some outflow component, they are very useful to determine a more exact BH mass. Their masses range over a few $10^7\ M_\odot$, and their column density is in the Compton-thick level. Their orientation is clearly random, position angles and inclinations are different form the large-scale disks. This explains why the radio jets, perpendicular to the accretion disk and torus, are frequently sweeping part of the large-scale gaseous disks of the galaxies. This enhances the coupling between jets and galaxies, and fosters the AGN feedback in the radio mode.

Dynamical simulations have been performed to check the stability of the molecular and dusty torus, especially in its expected thick shape. High-resolution 3D hydrodynamical simulations of a gas disk around a massive BH show that the gas disk remains very thin and stable in the very center, due to the stabilization of the BH Keplerian potential, but when the Toomre criterion is no longer satisfied, the disk become unstable to fragmentation and star formation. The supernovae feedback produces a large turbulence, and it is then possible to keep a flaring very thick disk, similar to the required obscuring torus (Wada & Norman 2002). However, the star formation rate must be exceptionally high to maintain the thickness (cf Figure 4.6). This is still true when taking into account the formation and cooling of the molecular hydrogen (Wada et al. 2009). The molecular fraction in the torus is of the order of 0.4, and the radiation field about 10 times that in the solar neighborhood. Taking into account the time-dependent chemistry in the X-ray dominated region (XDR) and the fountain effect of supernovae feedback, Wada et al. (2016) simulate the central outflow, in a bicone volume occupied by ionized diffuse gas. The dust is also entrained in the vertical direction, at the sublimation radius, in the inner parts of the thick molecular torus.

Clouds out of the clumpy torus are subject to strong radiation from the central engine, while they are inflowing towards the BH. The strong UV radiation penetrate only the surface of the cloud, before the heated dust re-radiates the energy in the infrared. Radiative transfer simulations have taken into account the dust grain

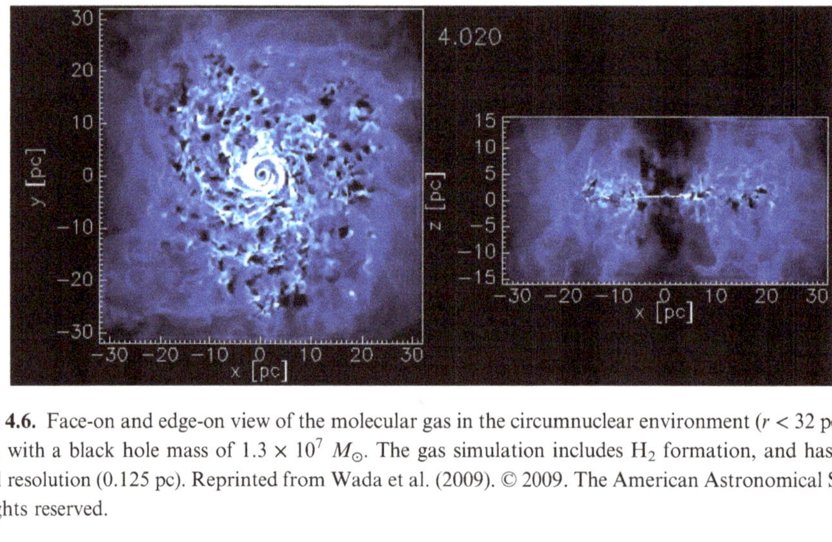

Figure 4.6. Face-on and edge-on view of the molecular gas in the circumnuclear environment (r < 32 pc) of an AGN, with a black hole mass of $1.3 \times 10^7\ M_\odot$. The gas simulation includes H_2 formation, and has a high spatial resolution (0.125 pc). Reprinted from Wada et al. (2009).

distribution, the heating of the gas from collisions with dust, the radial temperature gradient inside the cloud, the acceleration due to radiation pressure. While the shielded outer parts of the clouds still are inflowing by gravity, the inner parts are driven out and expanding by over-pressure, this creates shocks in the cloud, with instabilities. Through compression and stretching, the cloud is split in filaments, after several deformation phases. According to its density, part of the cloud will flow inwards, while another is pushed outwards by radiation pressure effects (Schartmann et al. 2011).

4.4 Polar Biconical Outflow

The AGN feedback has been observed for a long time, through emission lines revealing outflows of order 1000 km s^{-1}, called NLR to distinguish them from the BLR closer to the black hole (Veilleux & Osterbrock 1987). These outflows are resolved, extending up to 1 kpc sizes (Storchi-Bergmann et al. 1992). They are thought to be produced by the interaction of radiation from the accretion disk around the BH with the surrounding matter, and collimated by the thick doughnut like torus. The ionized gas outflows are often observed in X-ray and UV absorption (Crenshaw et al. 2003) in addition to optical emission lines, in nearby Seyferts, where they take the characteristic shape of bicone or hourglass (Antonucci & Miller 1985). The central engine is able to push the gas to a few hundreds of parsec, and is able to photo-ionize it at much larger distances, up to 20–30 kpc. At these distances, where the gas in the bicone is only emitting lines at rotational velocity, with a low dispersion, and is not outflowing, the emission-line region has been conventionally called the extended narrow line region (ENLR; Unger et al. 1987)

The two main mechanisms proposed to excite the emission lines are photoionization (e.g., Ferland & Osterbrock 1986) and shock ionization (e.g., Dopita & Sutherland 1995). Photoionization was already considered the dominant source of energy to produce the NLR emission lines just after the discovery of quasars (Greenstein & Schmidt 1964), and it was thought that radio jets and shocks would not have the right timescales and volumic extent. The small-scale structure of radio jets was not known however, and the detailed processes of the jet–ISM interaction, forming bow-shocks and cocoon (Bicknell et al. 1997). In the Seyfert 2 prototypical galaxies NGC 1068 and Mrk 3, most of the emission lines of the NLR could be produced by photoionization from the central source, however, it is also clear that there exists a tight radio jet interaction with the ambient medium, providing additional localized source of ionization through shocks. Evidence comes from velocity splitting of the gas at the radio knots, suggesting the expansion of a shell. These expanding gas shells are pushed by the cocoon of hot gas, shocked and heated by the radio jet (e.g., Capetti et al. 1999). One of the manifestations of these cocoon is that clouds are carried away laterally from the jet axis. The radio jet propagates in a clumpy and irregular medium, and is able to clear up a channel through the center of the bicone, which corresponds to its empty interior. This clearing might not be symmetric, since the ambient medium is not. In NGC 1068, where high spatial resolution data exist, the clearing is asymmetric, and the central AGN and torus appear in a lopsided configuration (Garcia-Burillo et al. 2019). There is also lateral expansion of clouds pushed aside by the cocoon (Das et al. 2006). Observing the X-ray hot gas wide extent in the NLR of Mrk 3, its flux ratio with optical emission lines, and the presence of shocks, Bogdan et al. (2017) conclude that the collisional excitation in shocks is important, and that the small-scale radio jets could play a significant role in Seyfert galaxies.

In NGC 1068, due to its proximity, detailed models have been confronted to observations. García-Burillo et al. (2014) found a rather massive molecular outflow, of 63 $M_{\odot}$/yr and concluded that the wind and radio jet are sweeping out the galactic plane, over a region extending from r ~50 pc out to r ~500 pc, due to the misalignment between the accretion disk and large-scale disk, and this is responsible for the high mass load of the molecular outflow.

Very detailed models of the NLR have been provided by Das et al. (2006) and Barbosa et al. (2014). There is evidence in the emission-line knots that the ionized gas is first accelerated up to a projected angular distance of 120 pc, and then decelerated, with a velocity approaching the systemic velocity at a projected distance of 300 pc. The deceleration of the clouds is likely due to collision with a clumpy and anistropically distributed ambient medium (Crenshaw & Kraemer 2000).

The schematic bicone model is represented in Figure 4.7. The general trends of radial velocities, first increasing, then decreasing with position is reproduces. In this model, the emitting material is flowing along the bicone axis, almost in the plane of the sky (inclined by 5° on the plane) explaining both redshifted and blueshifted velocities on both sides. The AGN wind occupies a hollow cone that is oriented along PA ~ 30° and is characterized by an opening angle of 80°, measured from the inner to the outer edge. The acceleration of the emission-line clouds is due to the radiation pressure and wind pressure. However, the model does not explain the lopsided position of the AGN, and the strong asymmetry between north–east and south–west, the outflow velocities being much higher to the north. This requires to take into account the additional push due to the radio lobe/cocoon shocks. Further

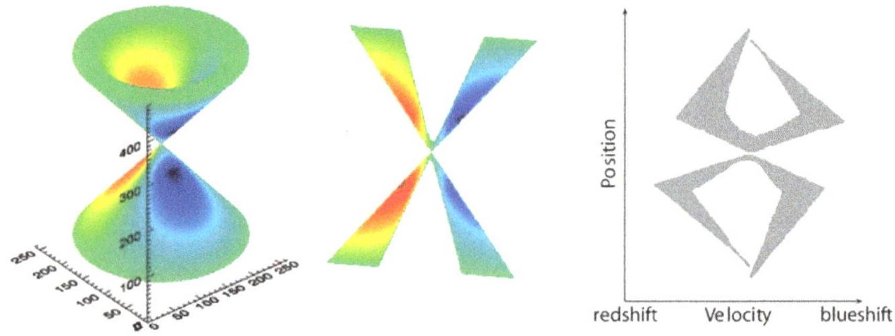

Figure 4.7. Left: bicone model of the ionized gas outflow, and rendering in 3D; the colors indicate the velocities. Middle: simulation of a spectral slit taken along a vertical in the plane of the sky, corresponding to cutting the 3D structure by a plane. The cone is hollow, and the cut yields two bands corresponding to the front (blueshifted) and the back (redshifted) of the cone. Right: corresponding position–velocity diagram of the slit. Reprinted from Das et al. (2006).

evidence showing the existence of a large-scale shock in the circumnuclear disk (CND) is provided by the comparison of the ALMA CO maps with the hourglass structure displayed by the NIR polarized emission imaged by Gratadour et al. (2015). This comparison, made by Garcia-Burillo et al. (2019) shows a striking correspondence between the inner edge of the CND and the edge-brightened arcs of NIR polarized emission, which are prominent at the northern and, most particularly, southern regions of the CND.

Barbosa et al. (2014) used the NIR emission lines, [Fe II], H_2–2μm, and Brγ. The warm H_2 is following the CND morphology of the cold CO-emitting gas with a characteristic empty ring in the SW, while the [Fe II] emission is the best tracer of the outflowing NLR, at v ~ 1000 km s^{-1}. It is tracing a wider region of outflowing gas, since it comes from a region where hydrogen is only partially ionized, excited by X-rays penetrating deeper than photons, and by shocks. This wider region is seen in Figure 4.8, and justifies that the outflow model is no longer a bicone, but an hourglass-shape or lemniscate. The Brγ region is much less extended, since coming from the fully ionized hydrogen region. The H_2 region is expanding in the CND plane with much lower velocities (~200 km s^{-1}), like the cold clouds.

The molecular outflow, detected at 50–300 pc scale in the direction of the radio jet, could be entrained by both the ionized gas wind or the radio jet cocoon. It is not easy to disentangle the mechanisms, since the molecular gas is flattened in the rotating molecular torus. It is possible to see, however, a slight vertical extension in the HCO+(4–3) emission of the torus. The AGN wind could peel off the inner layers of the torus and push the gas outward. This tendency is better seen in the continuum, showing the radiation pressure on dust (Impellizzeri et al. 2019).

Another example of radio jet sweeping out the ambient medium in the disk of a galaxy is the nearby lenticular Seyfert 2 IC5063. It has a strong radio jet of total size

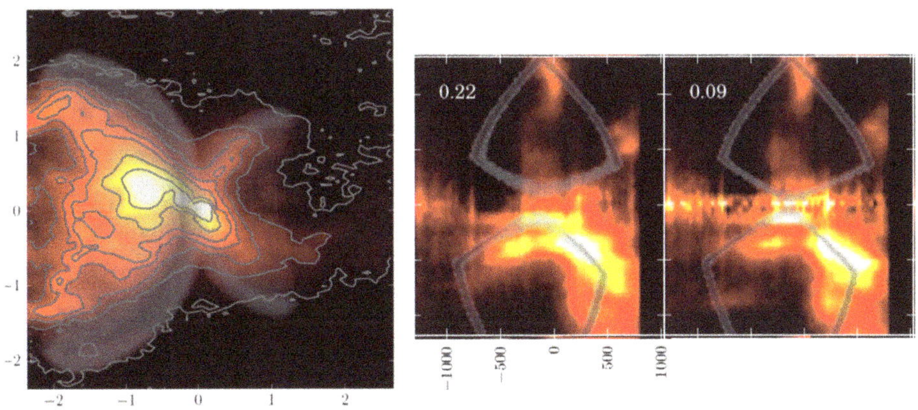

Figure 4.8. Left: the structure fitting best the [Fe II] image of NGC 1068 is not a bicone, but an hourglass-shape or lemniscate, here in gray shade, superposed on a color plus contours image of the [Fe II] emission. Right: two position–velocity diagrams, corresponding to two slits, parallel to the axis, and 0.22″ and 0.09″ offset from the center. The model is in gray shade. Reproduced from Barbosa et al. (2014). © 2014. The Royal Astronomical Society. All rights reserved.

about 1 kpc, corresponding to the NLR size of the optical emission lines, which are observed to be outflowing at velocities ~1000 km s^{-1}, in the vicinity of the radio knots. Both radio jet and NLR are co-spatial, aligned on the major axis of the disk (cf Figure 4.9). The HI-21cm line reveals a massive disk of atomic gas in emission (8.4×10^9 $M_\odot$), and a blueshifted, broad (700 km s^{-1}) absorption towards the nucleus, revealing a strong outflow (Morganti et al. 1998). The multiphase outflows, in ionized, atomic and molecular gas, appear to be pushed by strong shocks at the radio jet/cocoon interaction with the ambient medium. The UV radiation from the AGN is however responsible for the photoionization in a cone of the optical emission-line gas. In the NIR emission lines of [Fe II], Pα or H_2, a complex and distorted outflow morphology, with longitudinal and lateral outflows of the order of

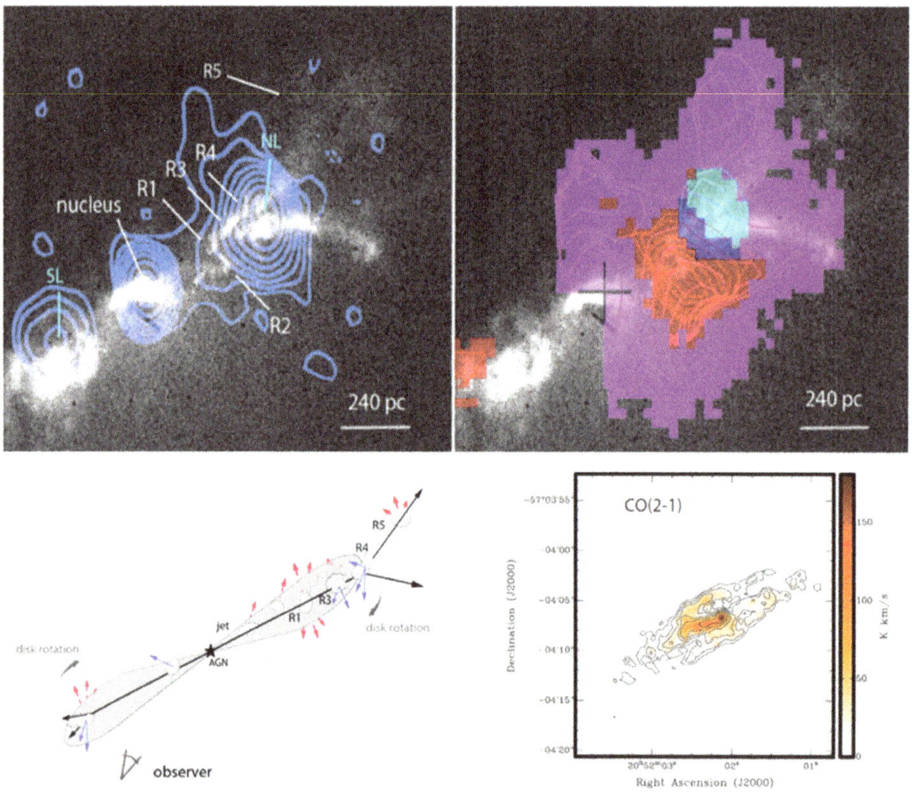

Figure 4.9 The NLR region in the Seyfert 2 galaxy IC5063. Top left: the 17.8 GHz radio jet contours (blue) are superposed on the HST narrow-band image containing Hβ and [O III]. Top right: on the same image are superposed the [Fe II] emission contours, and all the gas outflows shaded in cyan (blueshifted component), in red (redshifted one), and magenta for the intermediate velocities. Bottom left: schematic representation of the jet (full black line) and the cocoon (gray shade), with the gas outflow velocity indicating blue-shift, when the gas is in front, and redshift when remote, since it is pushed by the hot cocoon. Reprinted from Dasyra et al. (2015). © 2015. The American Astronomical Society. All rights reserved. Bottom right: ALMA map of the CO (2-1) emission, following the NLR direction. Reproduced with permission from Oosterloo et al. (2017). © ESO.

~500–1000 km s^{-1} show a good correlation with the radio jet knots (Dasyra et al. 2015).

In the nearby universe, the AGN are frequently of low-luminosity, and may not have a sufficiently high Eddington ratio to launch a powerful wind from their accretion disk. However, as described before, observations still require a dusty obscuring screen, to account for a large fraction of type 2 AGN, even if the doughnut tori of the unification paradigm are not observed. It has been proposed instead that the obscuration is provided by the dusty outflows, that have been seen in the mid-IR images of polar dust (Asmus et al. 2016). Since the radiation cross-section of IR radiation on dust is orders of magnitude greater than the Thomson cross-section on dust-free ionized gas, it is still possible to launch radiation-driven winds after the sublimation radius, when dust can survive (e.g., Figure 4.1). Radiative hydrodynamical simulations have been carried out to quantify this phenomenon, and in particular on clumpy media, schematized by clumps as particles, in order to be able to take into account the own radiation of the dust clouds (Venanzi et al. 2020). The strength of the wind obtained depends on the Eddington ratio and the column density of the dusty clumps, as shown in Figure 4.10. The comparison with the nearby Seyferts observed in the GATOS

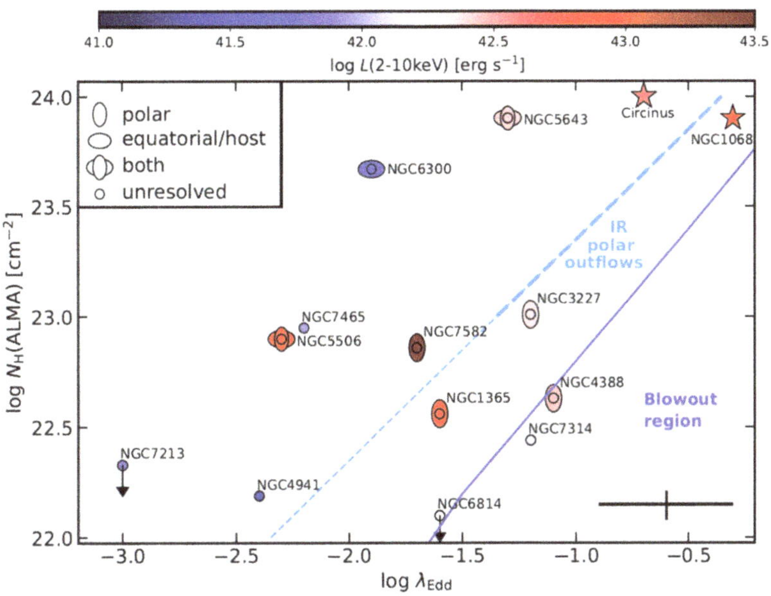

Figure 4.10. Study of a sub-sample of a dozen Seyfert from the GATOS project. Most of these galaxies have their Eddington ratio λ_{Edd} sufficient to produce a radiation-driven dust wind. However, some winds are not launched because of high gas column densities N_H, as shown in this diagram of N_H versus λ_{Edd}. The morphology of the torus-wind region is indicated in the top-left labels. The symbol color indicates the 2–10 keV luminosity. Below the full blue line, the whole nuclear region will be blown out (Fabian et al. 2008). The dashed blue line indicates the limit where the AGN radiation together with the IR radiation pressure from the dust clouds themselves balance gravity giving rise to polar dusty outflows (Venanzi et al. 2020). Reproduced with permission from Alonso-Herrero et al. (2021). © ESO.

(Galaxy Activity, Torus and Outflow Survey) project are in rough agreement with these predictions (Alonso-Herrero et al. 2021).

4.5 Misalignment

From the observations of molecular tori (e.g., Alonso-Herrero et al. 2018; Combes et al. 2019; Garcia-Burillo et al. 2021), it appears clearly that the tori are frequently misaligned and tilted with respect to the main disk. The observations are compatible with a random orientation. This is not unexpected, since other ways to determine the orientation of the accretion disk have come to the same conclusion. For instance, Fischer et al. (2013) studied NLR kinematics dominated by radial outflow, fitted them by biconical models, and derived the inclination of obscuring torus from the plane perpendicular to the bicone axis. From the clear signatures of biconical outflows in 17 individual AGN, they found no correlation between the inclination of the torus and the host galaxy. They find that the orientation of the torus on the line of sight for Seyfert 1 and 2 is compatible with the unification model.

There are several reasons why the nuclear disks decouple from the host galaxy disks. First the dynamical timescales are completely different, from below 0.5 Myr in the torus, to more than 200 Myr in the galaxy disk. The torus, inside the black hole's sphere of influence, is subject to a spherical potential, and does not keep for a long time information about the orientation and flattening of the rest of the galaxy. The torus is replenished constantly through accretion of gas from kpc scales in the disk, but from a disk thickness larger than the torus size. The angular momentum of the accreted gas may then vary. The more so as the star formation in the circumnuclear ring is active, and the supernovae feedback can eject gas above the plane, which will come back through the fountain effect after some interaction with hot gas in the halo. Collisions and merging with other clouds contribute to randomize the new orientation of accreted gas (e.g., Emsellem et al. 2015). This process is part of a renewed cycle, where the angular momentum of infalling gas is able to progressively change the BH spin, and reciprocally, the BH spin will align the gas in the accretion disk.

Given this constant re-orientation and warping of the torus, Lawrence & Elvis (2010) propose that the tilted disk, coming from the misalignment of the large-scale disk and molecular torus, provide most of the obscuration in type 2 AGN. In this model, the warped outer disk presents a wide solid angle to the inner disk and part of the UV optical radiation is absorbed and re-emitted in the IR. This enters in competition with the dusty outflow. However, it has not been shown in better agreement with the mid-IR or molecular observation.

Different mechanisms of misalignment between the nuclear and large-scale galaxy disks have been proposed. The first of them is the radiation-driven warping instability (Pringle 1996). An initially tilted optically thick disk, which absorbs the radiation from the central AGN, receives in each point some momentum from the radiation, but no torque, because of the radial direction of the photon received. But then it re-radiates in a direction normal to its surface, which is not radial if the disk is warped. This produces torques on the disk, which maintain and amplify the warping. Maloney & Begelman (1997) have developed the linear growth analytically, which depends on several parameters, in particular the viscosity α. Assuming the AGN luminosity is powered by accretion eliminates the unknown viscosity parameter. The instability occurs for radii $R > 0.1$ pc $M_{BH}/(10^8\ M_\odot)$. They have tested the efficiency of the mechanism through simulations, and showed that both retrograde and prograde precession can occur with respect to the disk rotation. A 3D rendering of the growth and evolution of this instability is shown in logarithmic radial scale in Figure 4.11. A second mechanism is the magnetic instability, and consequent torques but compatibility with AGN observations is contrived, it is more adapted to accretion disks around magnetic stars, white dwarfs or neutron stars (Pfeiffer & Lai 2004)

A third mechanism is due to the relativistic "frame-dragging" effect of a rotating black hole, the Lense–Thirring precession, leading to the Bardeen–Petterson (1975) effect. Let us assume that the accretion disk, replenished from the larger-scale material, has a random orientation with respect to the black hole spin (cf Figure 4.12). Through the relativistic frame-dragging effect, the black hole tends

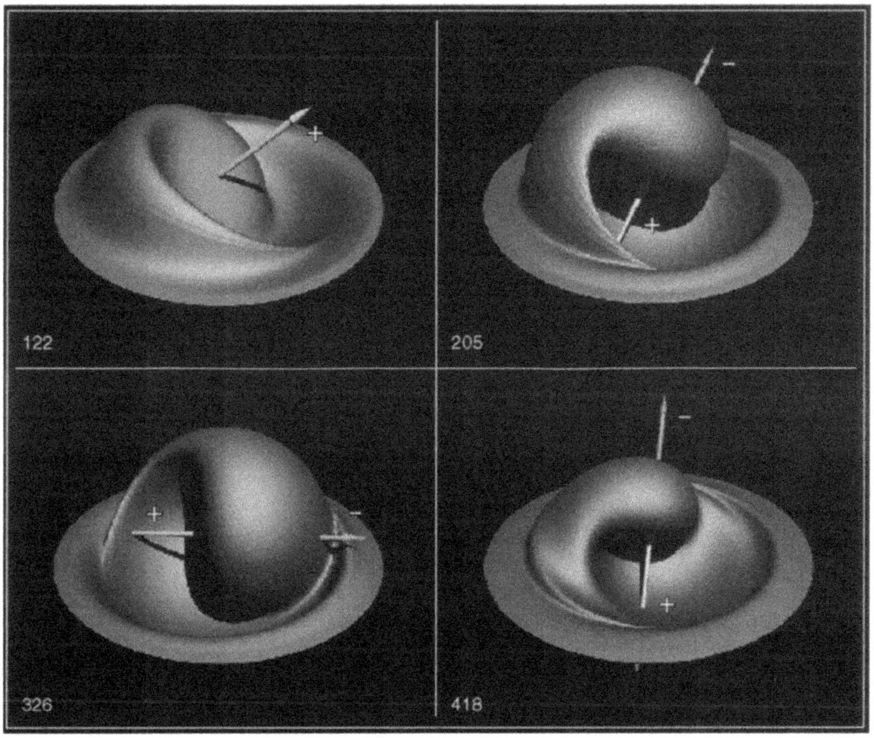

Figure 4.11. Simulations of the growth and non-linear evolution of radiation-induced warps, taking self-shadowing into account. The disk is shown at the dimensionless times 122, 205, 326 and 418. The radius is plotted in logarithmic scale, from log $r = 0$, to log $r = 4$, and the warp instability is allowed only to log $r = 3$ ($r = 10^3$, in dimensionless units). Reproduced from Pringle (1997).

to align the gas angular momentum to its spin direction. This produces torques which tend to align the inner parts of the accretion disk with the black hole equator. The inner and outer disk are no longer aligned. The disk develops a warping up to distances 10^2 to 10^4 Schwarzschild radius R_S. The precession of the disk and its warp can extend up to 1 pc ($M/10^9\ M_\odot$). Two regimes can be considered, according to the amplitude of viscosity. In case of high viscosity, the diffusion regime, the disk is warping smoothly and continuously, and is not teared off. This occurs when $\alpha > H/R$, where H is the height of the disk, and R its radius. In low viscosity, the regime of bending waves, when $\alpha < H/R$, the disk can break (Papaloizou & Pringle 1983). It then split in several rings, with different inclinations and processing rates. Then the various rings, with differential precession, collide, and drive the gas to fuel the AGN more quickly. The simulations by Nealon et al. (2015) have illustrated the two regimes (see Figure 4.12). The efficiency of the Bardeen–Petterson effect has been debated. Some works did not find the alignment of accretion disks with the black hole equator for all values of the parameters, in general relativistic magneto-hydrodynamic simulations (Zhuravlev et al. 2014; Banerjee et al. 2019). Polko & McKinney (2017) compare the Lense–Thirring precession and a Blandford–Znajek (1977) jet torque, in some sense the gravity and the electromagnetic mechanisms. In some cases, depending on the AGN efficiency, the Eddington ratio and the strength of the magnetic field, the electromagnetic process is more efficient. The observed jet precession will then indicate its provenance from the black hole or the accretion disk.

These warping instabilities are ubiquitous in the observation of the maser disks around black holes. Water masers in the prototypes NGC 4258 (Herrnstein et al. 1999) and NGC 1068 (Gallimore et al. 2004) are detected on 0.3–0.8 pc warped disks. The warping of the rotating disk is obvious in Figure 4.13 (Argon et al. 2007). These observations have been interpreted through the Lense–Thirring effect and/or

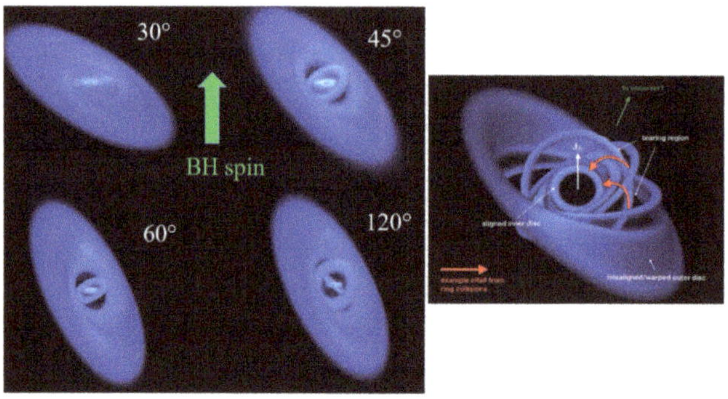

Figure 4.12. Simulation of the effect of black hole spin on warped or misaligned accretion disks. Left: initially the disks are misaligned with the black hole spin at 30°, 45°, 60°, and 120°. The spinning black hole, through the Lense–Thirring precession, tends to align the disk (Bardeen–Peterson effect). When the misalignment is larger than 45°, the disk fragments in discrete rings, which precess independently at different rates, which soon cancels out angular momentum and lead to rapid accretion. Reproduced from Nealon et al. (2015). Right: An extreme example of fragmentation, differential precession and rapid accretion of the gas. Reproduced from Pounds et al. (2018).

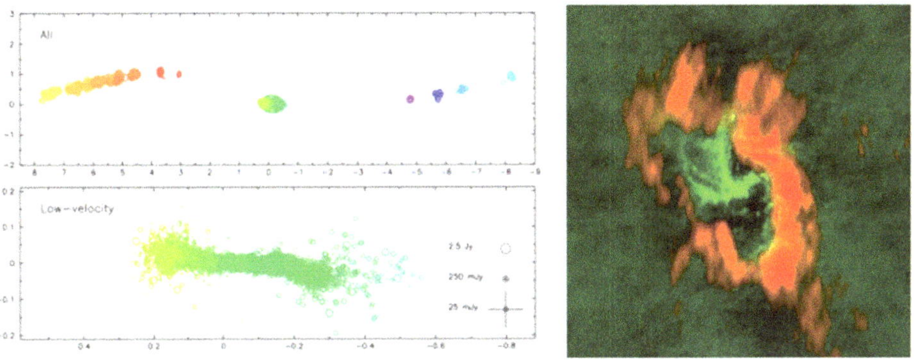

Figure 4.13. Misalignment and warps in circumnuclear disks around central black holes. Left: the H_2O maser disk of NGC 4258, with colors indicating the velocities (redshifted at East, blueshifted at West). The scale is indicated in milli-arcsec (1 mas = 0.035 pc). The whole disk is shown at top, and the center is zoomed in at the bottom. Reprinted from Argon et al. (2007). © 2007. The American Astronomical Society. All rights reserved. Right: the mini-spiral of ionized gas (3.6 cm continuum in green) and the HCN (1–0) line emission in the circumnuclear ring (in orange) in the center of the Milky Way. The box is 5.8 pc on a side. Reprinted from Yusef-Zadeh et al. (2008). © 2008. The American Astronomical Society. All rights reserved.

the radiation-driven warps. The fitting of the observations has been done successfully for NGC 4258 by Martin (2008), and for NGC 1068 by Caproni et al. (2006). In NGC 4258, the maser disk of ~0.2 pc in size is misaligned by 119° from the galaxy disk, the jet is in the plane of the galaxy, sweeping the interstellar medium over kpc scales In NGC 1068, the model based on the Bardeen–Petterson mechanism helps to quantify the timescale of evolution. The disk is aligned with the black hole spin, until the radius $R_{\rm BP} = 10^{-5}$ to 10^{-4} pc, then warps. The alignment timescale is 7580 yr, the misalignment angle 40° and the velocity of the jet $0.17c$. According to the mass swallowed, it is then possible to compute a timescale for the BH to re-arrange completely its spin, since it is reciprocally torqued by its successive accretion disks, and its angular momentum continuously evolves.

Another striking example of misalignment is in our own Galaxy, that we see edge-on. The circumnuclear pictures all look much less edge-on, as shown in Figure 4.13. There is a dense molecular circumnuclear ring (CNR) 2–3 pc in radius, traced by the HCN (1–0) line emission in orange, encircling a 3-arms mini-spiral of ionized gas in green (radio continuum emission). The apparent inclination is ~20° with respect to the Galactic plane; the mini-spiral has an estimated mass of 60 $M_\odot$ and the CNR 10^6 $M_\odot$.

Bibliography

Aalto, S., Martin, S., Costagliola, F., et al. 2015, A&A, 584, A42

Alonso-Herrero, A., Pereira-Santaella, M., Garcia-Burillo, S., et al. 2018, ApJ, 859, 144

Alonso-Herrero, A., Garcia-Burillo, S., Hoenig, S. F., et al. 2021, A&A, 652, A99

Antonucci, R. R. J., & Miller, J. S. 1985, ApJ, 297, 621

Argon, A. L., Greenhill, L. J., Reid, M. J., et al. 2007, ApJ, 659, 1040

Asmus, D., Hönig, S. F., & Gandhi, P. 2016, ApJ, 822, 109

Asmus, D. 2019, MNRAS, 489, 2177
Audibert, A., Combes, F., Garcia-Burillo, S., et al. 2019, A&A, 632, 33
Banerjee, S., Chakraborty, C., & Bhattacharyya, S. 2019, ApJ, 870, 95
Bannikova, E., Yu, A. V., Sergeyev, A. V., Akerman, N. A., et al. 2021, MNRAS, 503, 1459
Barbosa, F. K. B., Storchi-Bergmann, T., McGregor, P., et al. 2014, MNRAS, 445, 2353
Bardeen, J. M., & Petterson, J. A. 1975, ApJ, 195, L65
Bicknell, G. V., Dopita, M. A., & O'Dea, C. P. O. 1997, ApJ, 485, 112
Blandford, R. D., & Znajek, R. L. 1977, MNRAS, 179, 433
Bogdan, A., Kraft, R. P., Evans, D. A., et al. 2017, ApJ, 848, 61
Buta, R., Treuthardt, P. M., Byrd, G.-G., et al. 2000, AJ, 120, 1289
Capetti, A., Axon, D. J., Macchetto, F. D., et al. 1999, ApJ, 516, 187
Caproni, A., Abraham, Z., & Mosquera Cuesta, H. J. 2006, ApJ, 638, 120
Collin, S., & Zahn, J.-P. 2008, A&A, 477, 419
Combes, F., Garcia-Burillo, S., Audibert, A., et al. 2019, A&A, 623, A79
Crenshaw, D. M., & Kraemer, S. B. 2000, ApJ, 532, L101
Crenshaw, D. M., Kraemer, S. B., & George, I. M. 2003, ARA&A, 41, 117
Crocker, D. A., Baugus, P. D., & Buta, R. 1996, ApJS, 105, 353
Das, V., Crenshaw, D. M., Kraemer, S. B., et al. 2006, AJ, 132, 620
Dasyra, K. M., Bostrom, A. C., Combes, F., & Vlahakis, N. 2015, ApJ, 815, 34
Donmez, O. 2014, MNRAS, 438, 846
Donmez, O. 2017, MPLA, 32, 1750108
Dopita, M. A., & Sutherland, R. S. 1995, ApJ, 455, 468
Elitzur, M., & Shlosman, I. 2006, ApJ, 648, L101
Emsellem, E., Renaud, F., Bournaud, F., et al. 2015, MNRAS, 446, 2468
Fabian, A. C., Vasudevan, R. V., & Gandhi, P. 2008, MNRAS, 385, L43
Fazeli, N., Busch, G., Eckart, A., et al. 2020, A&A, 638, A53
Ferland, G. J., & Osterbrock, D. E. 1986, ApJ, 300, 658
Fischer, T. C., Crenshaw, D. M., Kraemer, S. B., & Schmitt, H. R. 2013, ApJS, 209, 1
Gallimore, J. F., Baum, S. A., & O'Dea, C. P. 2004, ApJ, 613, 794
García-Burillo, S., Combes, F., Usero, A., et al. 2014, A&A, 567, A125
Garcia-Burillo, S., Combes, F., Ramos Almeida, C., et al. 2016, ApJ, 823, L12
Garcia-Burillo, S., Combes, F., Ramos Almeida, C., et al. 2019, A&A, 632, A61
Garcia-Burillo, S., Alonso-Herrero, A., Ramos Almeida, C., et al. 2021, A&A, 652, A98
Gratadour, D., Rouan, D., Grosset, L., et al. 2015, A&A, 581, L8
Greenstein, J. L., & Schmidt, M. 1964, ApJ, 140, 1
Herrnstein, J. R., Moran, J. M., Greenhill, L. J., et al. 1999, Natur, 400, 539
Hönig, S. F. 2019, ApJ, 884, 171
Imanishi, M., Nguyen, D. D., Wada, K., et al. 2020, ApJ, 902, 99
Impellizzeri, C. M. V., Gallimore, J. F., Baum, S. A., et al. 2019, ApJ, 884, L28
Kamali, F., Henkel, C., Koyama, S., et al. 2019, A&A, 624, A42
Krolik, J. H., & Begelman, M. C. 1988, ApJ, 329, 702
Lawrence, A., & Elvis, M. 2010, ApJ, 714, 561
Maiolino, R., Risaliti, G., Salvati, M., et al. 2010, A&A, 517, A47
Maloney, P. R., & Begelman, M. C. 1997, ApJ, 491, L43
Martin, R. G. 2008, MNRAS, 387, 830
Morganti, R., Oosterloo, T., & Tsvetanov, Z. 1998, AJ, 115, 915

Nandra, K., O'Neill, P. M., George, I. M., & Reeves, J. N. 2007, MNRAS, 382, 194
Nealon, R., Price, D. J., & Nixon, C. J. 2015, MNRAS, 448, 1526
Nenkova, M., Sirocky, M. M., Ivezic, V., & Elitzur, M. 2008a, ApJ, 685, 147
Nenkova, M., Sirocky, M. M., Nikutta, R., et al. 2008b, ApJ, 685, 160
Netzer, H. 2015, ARA&A, 53, 365
Oosterloo, T., Raymond Oonk, J. B., Morganti, R., et al. 2017, A&A, 608, A38
Papaloizou, J. C. B., & Pringle, J. E. 1983, MNRAS, 202, 1181
Papaloizou, J. C. B., & Pringle, J. E. 1984, MNRAS, 208, 721
Peterson, B. M. 2001, Advanced Lectures on the Starburst-AGN Connection (Singapore: World Scientific), 3
Pfeiffer, H. P., & Lai, D. 2004, ApJ, 604, 766
Polko, P., & McKinney, J. C. 2017, MNRAS, 464, 2660
Pounds, K. A., Nixon, C. J., Lobban, A., & King, A. R. 2018, MNRAS, 481, 1832
Pringle, J. E. 1996, MNRAS, 281, 357
Pringle, J. E. 1997, MNRAS, 292, 136
Ramos Almeida, C., Alonso-Herrero, A., Levenson, N. A., et al. 2014, MNRAS, 439, 3847
Risaliti, G., Elvis, M., Fabbiano, G., et al. 2007, ApJ, 659, L111
Schartmann, M., Krause, M., & Burkert, A. 2011, MNRAS, 415, 741
Schawinski, K., Koss, M., Berney, S., et al. 2015, MNRAS, 451, 2517
Stalevski, M., Fritz, J., Baes, M., et al. 2012, MNRAS, 420, 2756
Storchi-Bergmann, T., Wilson, A. S., & Baldwin, J. A. 1992, ApJ, 396, 45
Tristram, K. R. W., Burtscher, L., Jaffe, W., et al. 2014, A&A, 563, A82
Unger, S. W., Pedlar, A., Axon, D. J., et al. 1987, MNRAS, 228, 671
Veilleux, S., & Osterbrock, D. E. 1987, ApJS, 63, 295
Venanzi, M., Hönig, S., & Williamson, D. 2020, ApJ, 900, 174
Viaene, S., Nersesian, A., Fritz, J., et al. 2020, A&A, 638, A150
Wada, K., & Norman, C. A. 2002, ApJ, 566, L21
Wada, K., Papadopoulos, P. P., & Spaans, M. 2009, ApJ, 702, 63
Wada, K., Schartmann, M., & Meijerink, R. 2016, ApJ, 828, L19
Yusef-Zadeh, F., Braatz, J., Wardle, M., et al. 2008, ApJ, 683, L147
Zhuravlev, V. V., Ivanov, P. B., Fragile, P. C., et al. 2014, ApJ, 796, 104

Chapter 5

Conclusion

This short chapter summarizes some of the highlights of the recent research in the domain of active galactic nuclei (AGN) feeding and feedback. There are several promising lines of research, where there remain unsolved issues, the main ones are underlined. Progress is expected in the next future with present instruments, and also with new telescopes, such as JWST and SKA.

5.1 Summary of Recent Knowledge and New Ideas

5.1.1 Molecular Outflows

One of the new aspects of galaxy evolution that has flourished in the last dozen years, due in particular to the Herschel satellite or millimeter interferometers such as ALMA and NOEMA, is the existence of molecular outflows, which sometimes are so massive that they could delay or quench star formation (e.g., Veilleux et al. 2020). Their mass loading factor can grow to a value of 10 times the star formation rate. Ionized gas outflows were known much earlier, but their mass and impact on the galaxies are estimated to be much smaller, for most of nearby galaxies (Fiore et al. 2017). Molecular outflows can be driven by supernovae feedback as well as AGN feedback, and it is sometimes hard to separate, since they occur frequently together in AGN host galaxies. In general, the maximum velocity reached by the outflow is low (~200 km s^{-1}) and insufficient to eject the gas completely out of the galaxy. The maximum velocity may help to separate supernovae form AGN feedback, since it is only by the latter that the escape velocity can be reached. The outflows could be energy-driven at their start, and turn to momentum-driven afterwards. Even with AGN, most of cases are not sufficiently strong to quench the star formation in galaxies, they are able to clear the central regions however (Fluetsch et al. 2019).

doi:10.1088/2514-3433/ac2a27ch5

5.1.2 The Polar Dust Distribution

The AGN unification paradigm relies on geometry and orientation on the line of sight of the circumnuclear regions around the supermassive black holes (BHs), to find universality behind the wide diversity of all AGN classes. In particular type 2 AGN are the same kind of objects as type 1 AGN, but obscured by a dusty torus, of doughnut shape. The latter has not been observed, in recent high resolution interferometric observations in the mid-infrared. A large progress has come from observations with the Very Large Telescope Interferometer (VLTI-MIDI) and the Keck of nearby Seyferts. The majority of the resolved images have revealed a polar dust geometry, and not an equatorial dusty torus (Tristram et al. 2014; Asmus et al. 2016; Asmus 2019). The model growing acceptance now is that presented in Chapter 4, where dust at the sublimation radius (of the order of 1 pc) is pushed and uplifted by IR radiation pressure in the edges of a hollow bicone, perpendicular to the accretion disk and molecular torus. This model has the advantage to solve the issue of the thickness of the doughnut-shape torus. In hydrodynamical simulations, it was not possible to keep the torus sufficiently thick for the required obscuration, or only when assuming strong supernovae feedback, as found in starbursts, to maintain its turbulence (Wada et al. 2009). The spectral energy distribution (SED) of the 1–10 pc scale dust around central black holes is now well modeled, and can help to identify the presence of AGN in a global galaxy spectrum (Viaene et al. 2020).

5.1.3 The Molecular Tori

The close environment of the central engines, and in particular within the sphere of influence of the black holes, has been reached in the sub-mllimeter domain by ALMA, revealing the molecular tori (Combes et al. 2019; Impellizzeri et al. 2019; Garcia-Burillo et al. 2021). In the majority of nearby Seyferts, a 10–20 pc radius molecular disk appears decoupled from the rest of the galaxy disk: it is decoupled by its morphology and kinematic implying a random orientation (position angle and inclination on the sky plane). This molecular torus is quite compact, with column densities to the Compton-thick level. Masses range between $1–3 \times 10^7\ M_\odot$. These disks are perpendicular to the radio jet, when it exists, and to the NLR outflowing gas, and therefore quasi aligned with the accretion disk. The molecular torus is sometimes located inside a trailing nuclear spiral, in particular in barred or double-barred galaxies (Audibert et al. 2019; Combes et al. 2019; Audibert et al. 2021). The sense of winding of these spirals is influenced by the gravity of the central black hole, and implies that the gas is then experiencing negative torques from the potential, and is falling inwards, to fuel the AGN

5.1.4 The X-ray Winds

Very close to the black hole, at a scale comparable to the accretion disk (up to 1000 Schwarzschild radii, or ~0.01 pc), ultra-fast outflows (UFO) have been discovered, with velocities as high as 0.3–0.4 c. These are detected in the Kα highly ionized iron line in absorption, blue-shifted to 7–10 keV (e.g. Tombesi et al. 2010). These absorptions are rather common in high luminosity radio-quiet AGN. Their variability indicates a distance implying that the wind is coming from the accretion disk. These fast outflows would correspond to the wind coming from the inner accreting disk, while the warm absorbers, seen in soft X-rays, with velocities ~1000 km s^{-1}, would come from the outer disk (Tombesi et al. 2013). All these absorbers are following the same continuous scaling relation, positively correlating the column density and ionization parameter with the outflow velocity. Given the high ionization of the fast outflows, they might be accelerated by radiation pressure via Compton scattering. At least in high luminosity AGN, this wind might be responsible of the molecular outflows at much larger distances.

5.1.5 AGN Feedback Efficiency in BCG

One of the best cases for AGN feedback efficiency is the cool core clusters and the feedback exerted in the kinetic or radio mode by the AGN in the brightest central galaxy (BCG). There is now a consensus that the strong cooling rates expected in a simplistic spherically symmetric model without feedback are reduced by at least an order of magnitude, when the regulating action of AGN heating is taken into account (e.g. Fabian 2012). The AGN radio jets push the thermal X-ray gas of the cluster, creating cavities, filled with a cocoon of diffuse plasma, progressing outwards from the center by buoyancy. Some low-entropy hot gas is uplifted by the bubble and reaches a zone where the cooling time is shorter than the free-fall time. The gas is cooling, through multi-phase components, down to cold molecular gas, mapped with ALMA, associated with warm Hα gas mapped with IFU like MUSE on the VLT (Olivares et al. 2019). Hydrodynamical simulations have shown that thermal instabilities occur as soon as the cooling time fall below 10 times the free-fall time, and chaotic cold gas accretion ensues (e.g., Gaspari et al. 2017). This scenario might be extended to massive galaxy groups, where an X-ray atmosphere has been detected.

5.2 Remaining Issues

5.2.1 Relative Role of Radiation-driven Winds and Radio Jets

There are two basic regimes for AGN feedback: the radiative mode, when the Eddington ratio is larger than a critical value (~0.01), creating a wind from the accretion disk, and a kinetic or radio mode at low Eddington ratios, when the radio jets are mechanically providing energy for the gas outflows. There could be a mixture of these two regimes, at relatively high Eddington ratios, and also due to the fact that the cross-section of IR radiation on dust is much higher than the Thomson cross-section on ionized gas. Then the dust beyond the sublimation radius could be radiation-driven at the edges of a hollow bicone, perpendicular to the accretion disk. One debated issue is to know the relative role of the radiative wind and radio jet in the ionized and molecular outflows. In many local Seyferts, the gas outflowing in the optical NLR is ionized by the AGN, but is also mechanically interacting with the radio jet and shocks of the radio cocoon. Some clear cases of jet–ISM interaction reveals knots and deviations in the radio jet, accompanied by secondary outflows in perpendicular or lateral directions (Dasyra et al. 2015; Bogdan et al. 2017). In some cases, the molecular outflow is so collimated that it can only have been entrained by a radio jet, even if it has faded or disappeared (Aalto et al. 2020).

5.2.2 (Re-)Formation of Molecular Clouds in Wind and Shock

The presence of dense molecular clouds in high velocity (up to 1000 km s^{-1}) outflows may rise problems: at these velocities, shocks will heat and destroy the molecules. Solutions were proposed that after crossing shocks and dissociation, gas instabilities re-form molecules *in situ* following gas fragmentation into clouds (Zubovas & King 2014). The multi-phase nature of the gas, fragmented in small clouds inside a more diffuse medium has been recognized to solve several problems, including the efficiency for a wind to entrain dense gas; dense clumps can be shredded and expand due to the wind shock (Hopkins & Elvis 2010), and then shatter again into small clouds after the shock (Sparre et al. 2019). These phenomena have not yet been probed in numerical simulations, since the small scale (sub-pc) fractal structure of clouds would require an unreachable spatial resolution. Pressure gradients could induce gas cooling, if the cloud cooling time falls below their destruction time (e.g. Gronke & Oh 2020). The formation of molecular hydrogen could then be very rapid (<1 Myr) in dense cold media by catalyst on grains. The (re)-formation of dust grains after a shock such as supernova shocks is possible (e.g. Draine 2003; Smith et al. 2009), but detailed physical processes, according to the grain structure have to be studied.

5.2.3 Black Hole Growth and Seeds at High Redshift

The black hole growth rate through accretion is proportional to its mass, and is thus very slow at the beginning, starting with stellar mass seeds. At any time, the Eddington luminosity imposes a ceiling in the accretion rate, and this limit is proportional to the BH mass. It is then a puzzle to explain the observation of a dozen quasars at $z = 6$–7 with masses larger than $10^9\ M_\odot$ (e.g. Venemans et al. 2016). It is important to understand why the $M-\sigma$ scaling relation breaks at high redshift, when the BH seems to grow faster than its bulge host. Several solutions have been proposed: either the BH in the early times accrete exceptionally several times faster than their Eddington rate, and this continuously, or the BH collapses directly from a massive cloud, due to primordial abundance, and lack of fragmentation. The seeds would then be much more massive than stellar, and the growth rate would be strong since the start. This however assumes contrived suppression of H_2 molecules formation, and is quite debated. The existence of intermediate mass black holes (IMBHs) are part of the puzzle, and these have not yet been unambiguously detected locally, although AGN signatures have been observed in dwarf galaxies (Chilingarian et al. 2018).

5.2.4 AGN Negative or Positive Feedback

The observed luminosity function of galaxies and its confrontation to the prediction of ΛCDM models require that AGN feedback has a strong negative effect to suppress the formation of large star forming galaxies. The AGN feedback has been shown to be efficient in galaxy clusters to moderate and regulate cooling flows. However, in nearby massive galaxies, the feedback appears insufficient to quench star formation. In some cases, positive feedback has been observed (e.g. Salomé et al. 2016; Maiolino et al. 2017). Simulations of AGN feedback have proposed both positive (Gaibler et al. 2012) or negative (Dubois et al. 2012) AGN feedback, although sometimes with very low efficiency (Gabor & Bournaud 2014). A further difficulty is that AGN and starburst are frequently triggered by the same external gas accretion. It is paramount to probe observationally the efficiency of AGN feedback at the best epoch of its action, i.e., at the peak ($z = 2$) of the cosmic black hole accretion rate (Madau & Dickinson 2014). These studies will gain a boost with the infrared observations of JWST. New emission line diagnostics will be used in the infrared to separate AGN and star formation (Kewley et al. 2019).

5.2.5 Unbiased Statistics of Outflows and ΛCDM Requirements

Our knowledge of molecular outflows, the most massive in capacity of depleting galaxies from their star formation fuel, and suppress over-massive galaxy formation, is still young, and has experienced a boost with ALMA and NOEMA. All samples that have been studied up to now are quite small, and biased to the most active in star formation and AGN, susceptible to host an outflow. The atomic cold phase, a very important ISM component playing the role of gas reservoir for star formation, participating in the outflow, has only been observed in emission with the HI-21cm line in very local galaxies. It has been established that the environment can strip this gas, leading to galaxy quenching (Boselli et al. 2014). In the future, the very sensitive SKA array will be able to explore HI emission across the cosmic time and the peak of cosmic star formation and black hole accretion rate. The HI outflow will also be widely studied through HI absorption (Morganti & Oosterloo 2018).

Bibliography

Aalto, S., Falstad, N., Muller, S., et al. 2020, A&A, 640, A104

Asmus, D., Hönig, S. F., & Gandhi, P. 2016, ApJ, 822, 109

Asmus, D. 2019, MNRAS, 489, 2177

Audibert, A., Combes, F., Garcia-Burillo, S., et al. 2019, A&A, 632, 33

Audibert, A., Combes, F., Garcia-Burillo, S., et al. 2021, A&A, arXiv: 2011.09133

Bogdan, A., Kraft, R. P., Evans, D. A., et al. 2017, ApJ, 848, 61

Boselli, A., Cortese, L., Boquien, M., et al. 2014, A&A, 564, A67

Chilingarian, I. V., Katkov, I. Y., Zolotukhin, I. Y., et al. 2018, ApJ, 863, 1

Combes, F., Garcia-Burillo, S., Audibert, A., et al. 2019, A&A, 623, A79

Dasyra, K. M., Bostrom, A. C., Combes, F., & Vlahakis, N. 2015, ApJ, 815, 34

Draine, B. T. 2003, ARA&A, 41, 241

Dubois, Y., Devriendt, J., Slyz, A., et al. 2012, MNRAS, 420, 2662

Fabian, A. C. 2012, ARA&A, 50, 455

Fiore, F., Feruglio, C., Shankar, F., et al. 2017, A&A, 601, A143

Fluetsch, A., Maiolino, R., Carniani, S., et al. 2019, MNRAS, 483, 4586

Gabor, J. M., & Bournaud, F. 2014, MNRAS, 441, 161

Gaibler, V., Khochfar, S., Krause, M., & Silk, J. 2012, MNRAS, 425, 438
Garcia-Burillo, S., Alonso-Herrero, A., Ramos Almeida, C., et al. 2021, A&A, 652, A98
Gaspari, M., Temi, P., & Brighenti, F. 2017, MNRAS, 466, 677
Gronke, M., & Oh, S. P. 2020, MNRAS, 492, 1970
Hopkins, P. F., & Elvis, M. 2010, MNRAS, 401, 7
Impellizzeri, C. M. V., Gallimore, J. F., Baum, S. A., et al. 2019, ApJ, 884, L28
Kewley, L. J., Nicholls, D. C., & Sutherland, R. S. 2019, ARA&A, 57, 511
Madau, P., & Dickinson, M. 2014, ARA&A, 52, 415
Maiolino, R., Russell, H. R., Fabian, A. C., et al. 2017, Natur, 544, 202
Morganti, R., & Oosterloo, T. 2018, A&ARV, 26, 4
Olivares, V., Salomé, P., Combes, F., et al. 2019, A&A, 631, A22
Salomé, Q., Salomé, P., Combes, F., & Hamer, S. 2016, A&A, 595, A65
Smith, N., Silverman, J. M., Chornock, R., et al. 2009, ApJ, 695, 1334
Sparre, M., Pfrommer, C., & Vogelsberger, M. 2019, MNRAS, 482, 5401
Tombesi, F., Cappi, M., Reeves, J. N., et al. 2010, A&A, 521, A57
Tombesi, F., Cappi, M., Reeves, J. N., et al. 2013, MNRAS, 430, 1102
Tristram, K. R. W., Burtscher, L., Jaffe, W., et al. 2014, A&A, 563, A82
Veilleux, S., Maiolino, R., Bolatto, A. D., & Aalto, S. 2020, A&ARV, 28, 2
Venemans, B. P., Walter, F., Zschaechner, L., et al. 2016, ApJ, 816, 37
Viaene, S., Nersesian, A., Fritz, J., et al. 2020, A&A, 638, A150
Wada, K., Papadopoulos, P. P., & Spaans, M. 2009, ApJ, 702, 63
Zubovas, K., & King, A. R. 2014, MNRAS, 439, 400

www.ingramcontent.com/pod-product-compliance
Ingram Content Group UK Ltd.
Pitfield, Milton Keynes, MK11 3LW, UK
UKHW060959120926
8556IPUK00033B/908

9 780750 330336